**Brian D. Harper**
*Ohio State University*

# Solving Dynamics Problems in Maple to Accompany Engineering Mechanics Dynamics

## Sixth Edition

### J.L. Meriam ~ L.G. Kraige

*Virginia Polytechnic Institute & State University*

John Wiley & Sons, Inc.

Cover Photo: © Courtesy of NASA/JPL

Bicentennial Logo Design: Richard J. Pacifico

To order books or for customer service please, call 1-800-CALL WILEY (225-5945).

ISBN-13   978- 0-470-09920-9

10 9 8 7 6 5 4 3 2 1

Printed and bound by Lightning Source

**THE WILEY BICENTENNIAL–KNOWLEDGE FOR GENERATIONS**

Each generation has its unique needs and aspirations. When Charles Wiley first opened his small printing shop in lower Manhattan in 1807, it was a generation of boundless potential searching for an identity. And we were there, helping to define a new American literary tradition. Over half a century later, in the midst of the Second Industrial Revolution, it was a generation focused on building the future. Once again, we were there, supplying the critical scientific, technical, and engineering knowledge that helped frame the world. Throughout the 20th Century, and into the new millennium, nations began to reach out beyond their own borders and a new international community was born. Wiley was there, expanding its operations around the world to enable a global exchange of ideas, opinions, and know-how.

For 200 years, Wiley has been an integral part of each generation's journey, enabling the flow of information and understanding necessary to meet their needs and fulfill their aspirations. Today, bold new technologies are changing the way we live and learn. Wiley will be there, providing you the must-have knowledge you need to imagine new worlds, new possibilities, and new opportunities.

Generations come and go, but you can always count on Wiley to provide you the knowledge you need, when and where you need it!

**WILLIAM J. PESCE**
PRESIDENT AND CHIEF EXECUTIVE OFFICER

**PETER BOOTH WILEY**
CHAIRMAN OF THE BOARD

# CONTENTS

# INTRODUCTION

Computers and software have had a tremendous impact upon engineering education over the past several years and most engineering schools now incorporate computational software such as Maple in their curriculum. Since you have this supplement the chances are pretty good that you are already aware of this and will have to learn to use Maple as part of a Dynamics course. The purpose of this supplement is to help you do just that.

There seems to be some disagreement among engineering educators regarding how computers should be used in an engineering course such as Dynamics. I will use this as an opportunity to give my own philosophy along with a little advice. In trying to master the fundamentals of Dynamics there is no substitute for hard work. The old fashioned taking of pencil to paper, drawing free body and mass acceleration diagrams, struggling with equations of motion and kinematic relations, etc. is still essential to grasping the fundamentals of Dynamics. A sophisticated computational program is not going to help you to understand the fundamentals. For this reason, my advice is to use the computer only when required to do so. Most of your homework can and should be done without a computer. A possible exception might be using Maple's symbolic algebra capabilities to check some messy calculations.

The problems in this booklet are based upon problems taken from your text. The problems are slightly modified since most of the problems in your book do not require a computer for the reasons discussed in the last paragraph. One of the most important uses of the computer in studying Mechanics is the convenience and relative simplicity of conducting parametric studies. A parametric study seeks to understand the effect of one or more variables (parameters) upon a general solution. This is in contrast to a typical homework problem where you generally want to find one solution to a problem under some specified conditions. For example, in a typical homework problem you might be asked something about the trajectory of a particle launched at an angle of 30 degrees from the horizontal with an initial speed of 30 ft/sec. In a parametric study of the same problem you might typically find the trajectory as a function of two parameters, the launch angle $\theta$ and initial speed $v$. You might then be asked to plot the trajectory for different launch angles and speeds. A plot of this type is very beneficial in visualizing the general solution to a problem over a broad range of variables as opposed to a single case.

As you will see, it is not uncommon to find Mechanics problems that yield equations that cannot be solved exactly. These problems require a numerical approach that is greatly simplified by computational software such as Maple. Although numerical solutions are extremely easy to obtain in Maple this is still the method of last resort. Chapter 1 will illustrate several methods for obtaining symbolic (exact) solutions to problems. These methods should always be tried first. Only when these fail should you generate a numerical approximation.

Many students encounter some difficulties the first time they try to use a computer as an aid to solving a problem. In many cases they are expecting that they have to do something fundamentally different. It is very important to understand that there is no fundamental difference in the way that you would formulate computer problems as opposed to a regular homework problem. Each problem in this booklet has a problem formulation section prior to the solution. As you work through the problems be sure to note that there is nothing peculiar about the way the problems are formulated. You will see free-body and mass acceleration diagrams, kinematic equations etc. just like you would normally write. The main difference is that most of the problems will be parametric studies as discussed above. In a parametric study you will have at least one and possibly more parameters or variables that are left undefined during the formulation. For example, you might have a general angle $\theta$ as opposed to a specific angle of 20°. If it helps, you can "pretend" that the variable is some specific number while you are formulating a problem.

This supplement has eight chapters. The first chapter contains a brief introduction to Maple. If you already have some familiarity with Maple you can skip this chapter. Although the first chapter is relatively brief it does introduce all the methods that will be used later in the book and assumes no prior knowledge of Maple. Chapters 2 through 8 contain computer problems taken from chapters 2 through 8 of your textbook. Thus, if you would like to see some computer problems involving the kinetics of particles you can look at the problems in chapter 3 of this supplement. Each chapter will have a short introduction that summarizes the types of problems and computational methods used. This would be the ideal place to look if you are interested in finding examples of how to use specific functions, operations etc.

This supplement uses Maple 10. Maple is a product Waterloo Maple, Inc., 450 Phillips Street, Waterloo, Ontario, Canada.

# AN INTRODUCTION TO MAPLE

1

This chapter provides an introduction to the Maple programming language. Although the chapter is introductory in nature it will cover everything needed to solve the computer problems in this booklet.

## 1.1 Numerical and Symbolic Calculations

A command line in a Maple worksheet contains a set of instructions followed by a semicolon or colon. Pressing enter causes Maple to process the line. If the line ends in a semicolon Maple will show the output on the next line. If the line ends in a colon Maple will still process the line but the output will not be shown. It is a good idea to always use a semicolon when you are writing and debugging a worksheet. Later, if you wish, you can change the line to end with a colon to hide superfluous or lengthy output. There are several places in this supplement where the output will be hidden by using a colon. This is done either for trivial assignments where output is not necessary or for those rare occasions where the output may be several lines or even pages.

In an actual Maple session, command lines will begin with "[>" and will be red. Output is centered and blue. In this supplement, command lines will begin only with ">" while output lines will follow on the next line and be indented. The order in which command lines are executed is important in Maple, however, the order in which they are written is not. For this reason you can execute a line once and then go back to the same line later and execute it again, perhaps getting a different result. Thus, it is almost inevitable that something will go wrong as you debug a program. If you get some unexpected result the first thing you should try is re-executing your worksheet. To this end it is highly advisable to start every session with the *restart* command. Then, if you need to have the program re-execute from the beginning you can select *Edit...Execute...Worksheet*.

Here are several examples illustrating typical numerical calculations.

```
> restart; # always start with restart
> 2+7-4;
```

$$5$$

```
> 5*12-4*3;
```
$$48$$

```
> 8^2;
```
$$64$$

```
> 3/10+12/4;
```
$$\frac{33}{10}$$

```
> sqrt(2);
```
$$\sqrt{2}$$

Note in the last two examples that Maple will not show a decimal representation of a number unless asked to do so. Converting a number to a decimal approximation is accomplished with the *evalf* function.

```
> evalf(sqrt(2));
```
$$1.414213562$$

By default, the number of significant digits is 10. You can specify the number of significant digits in the *evalf* command as follows.

```
> evalf(sqrt(2),5);
```
$$1.4142$$

You can also change the default with the *Digits* variable.

```
> Digits:=20:
> evalf(sqrt(2));
```
$$1.4142135623730950488$$

Later we will find that the *evalf* function can be used as an easy way to perform numerical integration.

In the previous examples we have been using Maple as if it were an expensive calculator. Maple is, however, much more than a calculator. It can, for example, perform algebraic manipulations upon variables as well as numbers. Such manipulations are generally referred to as symbolic calculations or computer algebra. Following are several simple examples of symbolic calculations.

```
> (a*x+b*x^2)/x^3;
```
$$\frac{a\,x + b\,x^2}{x^3}$$

```
> %*x^2;
```

$$\frac{a\,x + b\,x^2}{x}$$

Note that % is a shortcut for referring to the last expression evaluated by Maple. Similarly, %% would refer to the next to last expression and %%% to the one before that. Be sure you understand the following calculation.

> %*%%;

$$\frac{\left(a\,x + b\,x^2\right)^2}{x^4}$$

Sometimes Maple does not produce the simplest form of a symbolic operation. If you suspect this might be the case, try simplifying with the *simplify* command.

> simplify(%);

$$\frac{(a + b\,x)^2}{x^2}$$

Here is another example of the *simplify* command.

> (x^2+x^4)/(x^3);

$$\frac{x^2 + x^4}{x^3}$$

> simplify((x^2+x^4)/(x^3));

$$\frac{1 + x^2}{x}$$

## 1.2 Assignments, Names and Variables

The assignment command ":=" (two keystrokes, colon followed by equal sign) can be used to assign an object (usually a number or an expression) to a variable.

> restart;
> x := 3;

$$x := 3$$

> We := Love +Dynamics;

$$We := Love + Dynamics$$

In the first example, the number 3 is assigned to the variable *x*. In the second, the sum of two variables is assigned to a third variable *We*. Note that Maple is case sensitive. *We, we, wE and WE* are all different variables. An assigned variable is sometimes referred to as a name or an alias. Whenever Maple encounters *We* it will automatically substitute *Love + Dynamics*. Thus, it is also useful to think of assignments (names, aliases) as abbreviations. "*We*" is an abbreviation for "*Love*

+ *Dynamics*". To illustrate this feature, consider the following symbolic calculations.

> We^2;
$$(Love + Dynamics)^2$$

> 10^We;
$$10^{(Love + Dynamics)}$$

> We/We;
$$1$$

> Love:=9: Dynamics:= 16:
> sqrt(We);
$$5$$

It is important to understand the difference between the assignment operator (:=) and the equality operator (=). := assigns an object to a variable while = defines equalities. We'll use the equality operator primarily to write equations that we subsequently ask Maple to solve. This will be discussed in greater detail later. Here we will just give a simple example containing both operators.

> eqn := x^2 + y^2 = z^2;
$$eqn := 9 + y^2 = z^2$$

Here we have assigned an equality to the variable *eqn*. You may recall that $x$ has been previously assigned the value 3. Maple remembers this and automatically substitutes 3 for $x$. The *unassign* command removes a previous assignment.

> x;
$$3$$

> unassign('x'); # note that the variable must be placed in quotations.
> x;
$$x$$

Listing x before and after the *unassign* command just reinforces the change in assignment and is not necessary. It is important to understand at this point that the assignment for *eqn* has not changed even though the variable $x$ is no longer assigned. To see this let's list the variable *eqn*.

> eqn;
$$9 + y^2 = z^2$$

The reason for this is that the variable *eqn* was assigned when $x$ was equal to 3. Thus, unassigning $x$ has no effect on that previous assignment. If you want the name *eqn* to contain the variable $x$ then $x$ should be unassigned before *eqn* is assigned. This point is illustrated by the following.

> eqn := x^2 + y^2 = z^2;
$$eqn := x^2 + y^2 = z^2$$

> x:=3:
> eqn;
$$9 + y^2 = z^2$$

> unassign('x');
> eqn;
$$x^2 + y^2 = z^2$$

Now we see that unassigning $x$ recovers the original definition of *eqn*.

## 1.3 Functions

Maple has many basic mathematical functions built in. The following table provides a summary of those most useful in Dynamics.

Table 1.1 Built in mathematical functions.

| sin, cos, tan, cot, sec, etc. | Trig functions, Sine, Cosine etc. |
|---|---|
| arcsin, arccos, arctan, arcsec, etc. | Inverse trig functions |
| ^ | Raising to a power, $x\text{^}y = x^y$. |
| exp | Exponential function. $\exp(y) = e^y$. |
| log | Natural logarithm. |
| Log10 | Base 10 logarithm. |
| sqrt | Square root, $sqrt(x) = \sqrt{x}$ |

When using trig functions, be sure to remember that Maple uses radians by default. Also, the number $\pi$ is represented in Maple by Pi while pi is just the greek letter $\pi$. As the following lines show, it is not always easy to tell one from the other in Maple output.

> restart;
> Pi; pi;
$$\pi$$

$$\pi$$

Using Pi and pi as arguments in a function makes it clear that the first is a number while the second is a variable.

> sin(Pi/2); sin(pi/2);

$$1$$

$$\sin\left(\frac{1}{2}\pi\right)$$

All the greek letters (with the exception of the protected names Pi and gamma) can be used by spelling out their names. Type ?greek at a command prompt for a review of the greek letters and their spelling. Here are a few examples.

> cos(theta)+sin(Theta); tan(omega)-cot(Omega);

$$\cos(\theta) + \sin(\Theta)$$

$$\tan(\omega) - \cot(\Omega)$$

It turns out that the greek letter gamma ($\gamma$) is commonly used in Dynamics problems to denote an angle. If you do not need the gamma function (and you probably won't), feel free to unprotect it and then use it in your problems.

> gamma:=Pi/4;
*Error, attempting to assign to `gamma` which is protected*

> unprotect('gamma');
> gamma:=Pi/4;

$$\gamma := \frac{1}{4}\pi$$

Maple also allows the user to define functions. Here are two examples.

> f:= x ->x^3;

$$f := x \rightarrow x^3$$

> g:= (x,y) -> x^2+y^2;

$$g := (x,y) \rightarrow x^2 + y^2$$

The syntax for defining a function starts with the usual assignment operator ":=" followed by a list of variables and then what looks like an arrow but is really two key strokes, a minus sign "-" and a greater than symbol ">". Note that more than one variable requires the use of parentheses. User defined functions operate exactly like built in functions. Their arguments may be numbers, variables or expressions.

> f(3);

$$27$$

> f(2)+g(1,2)/g(3,1);

$$\frac{17}{2}$$

\> f(g(2,4));
$$8000$$

\> g(duck,goose);
$$duck^2 + goose^2$$

\> f(g(r^2,sqrt(r)));
$$\left(r^4 + r\right)^3$$

It is important to understand that there is nothing special about the choice of variables used to define a function. The following lines show that the structure of the function is not changed even if x and y are assigned.

\> x:=3: y:=3:
\> g(x,y);
$$18$$

\> g(m,q);
$$m^2 + q^2$$

# 1.4 Graphics

One of the most useful things about a computational software package such as Maple is the ability to easily create graphs of functions. As we will see, these graphs allow one to gain a lot of insight into a problem by observing how a solution changes as some parameter (the magnitude of a load, an angle, a dimension etc.) is varied. This is so important that practically every problem in this supplement will contain at least one plot. By the time you have finished reading this supplement you should be very proficient at plotting in Maple. This section will introduce you to the basics of plotting in Maple.

### Simple Plots of One or More Expressions

Simple plots are easily obtained with the *plot* command. The simplest plot call requires only the expression to be plotted and the range over which it is to be plotted. There are many advanced features of the plot command allowing graphs to be formatted in various ways. We will show the main options primarily by way of example. For more details, type ?plot, ?plot[structure], or ?plot[options] at any Maple command prompt ([>).

\> restart; # always start with restart
\> plot(5*x^2-0.5*x^3, x=0..10, color=black);

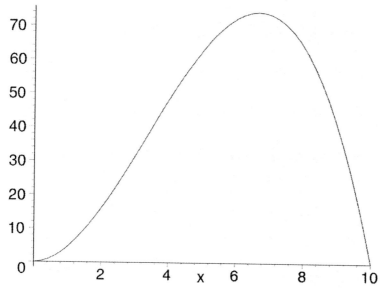

```
> f:=x^2*sin(x);
```
$$f := x^2 \sin(x)$$

```
> plot(f, x=-2*Pi..2*Pi, title=`x^2sin(x)`);
```

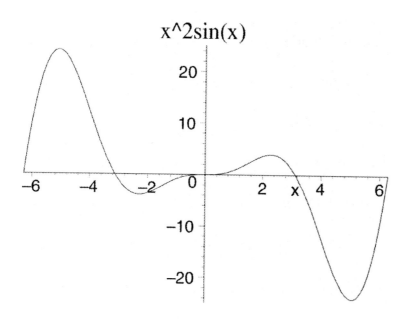

Plotting more than one function on a single graph is accomplished by replacing the single function with a list of functions in brackets [].

> g:=sqrt(x)*exp(-x); h:=x^2*10^(-x/2);

$$g := \sqrt{x}\ e^{(-x)}$$

$$h := x^2\ 10^{(-1/2\,x)}$$

> plot([g,h],x=0..5, color=black);

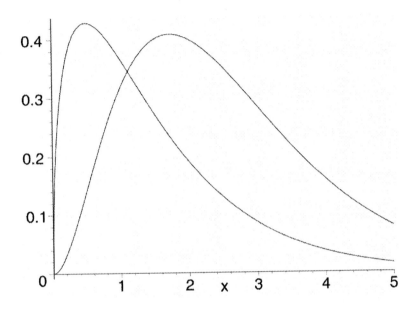

In an actual Maple session the two curves above would be plotted in different colors. If you want to distinguish results by color, its a good idea to use the color option. For example, the plot command plot([g,h],x=0..5, color=[red,blue]) would yield a graph where g is plotted red and h blue.

It is also possible, though somewhat tedious, to label individual curves in a plot. This requires the *display* and *textplot* procedures which must be loaded from the plots package with the command with(plots).

> with(plots):
> p1:=plot([g,h],x=0..5, color=black):
> t1:=textplot([1,0.41,"g"],color=black,font=[TIMES,ROMAN,16]):
> t2:=textplot([2.3,0.41,"h"],color=black,font=[TIMES,ROMAN,16]):

The parameters for *textplot* are the text and the coordinates at which it is to be plotted. Note that the individual *plot* and *textplot* structures are assigned names. These names are then input to the *display* procedure.

> display(p1,t1,t2);

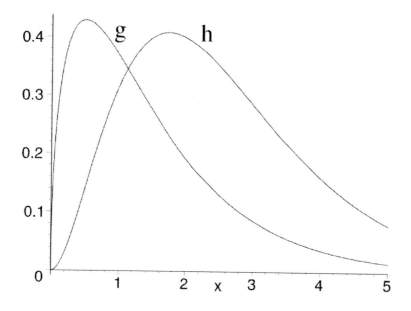

### Multiple Plots of a Single Function (Parametric Studies)

One of the most important uses of the computer in studying Dynamics is the convenience and relative simplicity of conducting parametric studies (not to be confused with parametric plotting). A parametric study seeks to understand the effect of one or more variables (parameters) upon a general solution. This is in contrast to a typical homework problem where you generally want to find one solution to a problem under some specified conditions. For example, in a typical homework problem you might be asked to find the reactions at the supports of a structure with a concentrated force of magnitude 200 lb that is oriented at an angle of 30 degrees from the horizontal. In a parametric study of the same problem you might typically find the reactions as a function of two parameters, the magnitude of the force and its orientation. You might then be asked to plot the reactions as a function of the magnitude of the force for several different orientations. A plot of this type is very beneficial in visualizing the general solution to a problem over a broad range of variables as opposed to a single case.

Parametric studies generally require making multiple plots of the same function with different values of a particular parameter in the function. Following is a very simple example.

```
> restart;
> f:=5+x-5*x^2+a*x^3;
```
$$f := 5 + x - 5\,x^2 + a\,x^3$$

What we would like to do is gain some understanding of how f varies with both x and a. It might be tempting to make a three dimensional plot in a case like this. Such a plot can, in some cases, be very useful. Usually, however, it is too difficult to interpret. This is illustrated by the following three dimensional plot of f versus x and a.

> plot3d(f,x=-10..10,a=-3..3,color=grey);

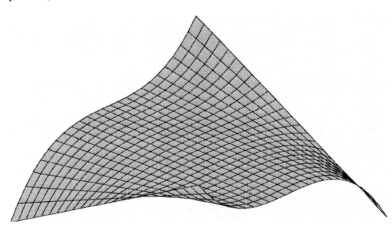

The plot above certainly is interesting but, as mentioned above, not very easy to interpret. In most cases it is much better to plot the function several times (with different values of the parameter of interest) on a single two dimensional graph. We will illustrate this by plotting f as a function of x for four values of the parameter a. There are a number of ways this can be done. Below are two methods: (a) re-assignment and (b) substitution.

> a:=-2: f1:=f: a:=-1: f2:=f: a:=1: f3:=f: a:=2: f4:=f:  # method (a)
> plot([f1,f2,f3,f4],x=-10..10,color=black);

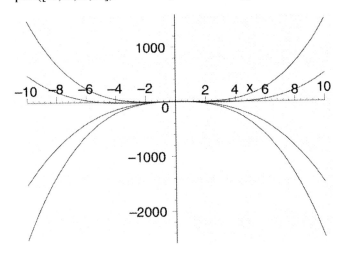

> unassign('a');
> plot([subs(a=-2,f),subs(a=-1,f),subs(a=1,f),subs(a=2,f)],x=-10..10,color=black); # method (b)

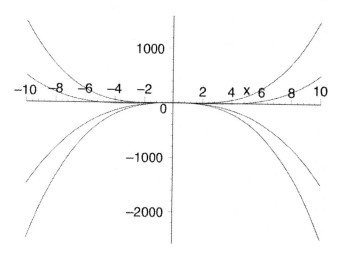

It is almost impossible to determine which curve corresponds to which value of a in the above two plots. This distinction can be made either by color coding or labeling, as discussed earlier.

### Parametric Plots

It often happens that one needs to plot some function y versus x but y is not known explicitly as a function of x. For example, suppose you know the x and y coordinates of a particle as a function of time but want to plot the trajectory of the particle, i.e. you want to plot the y coordinate of the particle versus the x coordinate. Since both x and y are known in terms of another parameter (in this case time t), it is possible to obtain the desired plot using a parametric plot in Maple. The calling sequence for the parametric plot is plot([x(t),y(t),t=range of t],h,v,options), which results in y being plotted versus x for values of the parameter t in the range specified. h and v specify the horizontal and vertical ranges for the plot.

Now let's consider a simple example with two functions f and g expressed in terms of a parameter t.

> f:=10*t*(2-t); g:=sin(5*t);
$$f := 10\,t\,(2-t)$$

$$g := \sin(5\,t)$$

To plot f versus g:
> plot([g, f, t=0..2],color=black,labels=['g','f']);

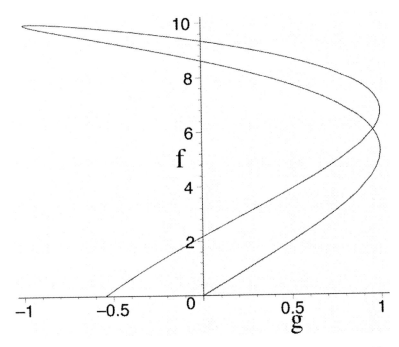

To plot several parametric plots on the same graph one should use the display procedure. Recall that this procedure must be loaded by typing with(plots): As an example, consider the following expressions for x and y in terms of a common parameter t:

```
> with(plots):
> y:=a*t-b*t^2;x:=c*sin(beta*t);
```

$$y := a\,t - b\,t^2$$

$$x := c\,\sin(\beta\,t)$$

Now suppose we want to specify that a=1, b=0.5, c=2 and then plot y versus x for three values of β (0.1, 0.25, 0.5). This is accomplished as follows.

```
> a:=1: b:=0.5: c:=2:
> beta:=0.1:p1:=plot([x, y, t=0..2],0..2,0..0.6,color=black):
> beta:=0.25:p2:=plot([x, y, t=0..2],0..2,0..0.6,color=black):
> beta:=0.5:p3:=plot([x, y, t=0..2],0..2,0..0.6,color=black):
> display([p1,p2,p3]);
```

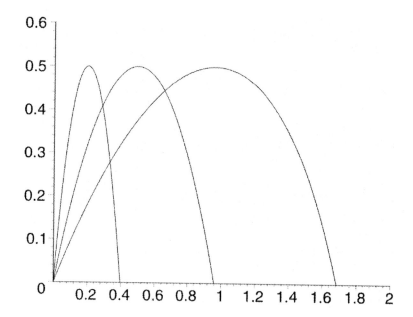

## 1.5 Differentiation and Integration

Mechanics problems often require integration and differentiation. In Maple, you can perform these operations either numerically or symbolically. In this supplement we will generally find symbolic results whenever possible and leave numerical solutions as a method of last resort.

```
> restart;
> diff(x*tan(x^3),x);
```
$$\tan(x^3) + 3\,x^3\,(1 + \tan(x^3)^2)$$

```
> f:=a*sec(b*t);
```
$$f := a\,\sec(b\,t)$$

```
> diff(f,t);
```
$$a\,\sec(b\,t)\,\tan(b\,t)\,b$$

Higher order derivatives can be obtained in two ways as illustrated by the following.

```
> g:=a*log(b*x^2);
```
$$g := a\,\ln(b\,x^2)$$

```
> diff(g,x,x,x); diff(g,x$3); # each expression evaluates the third derivative of g.
```

$$4 \frac{a}{x^3}$$

$$4 \frac{a}{x^3}$$

You can also use derivatives in defining functions. As an example, suppose a particle moves in a straight line and its position s is known as a function of time. From your elementary physics course you probably know that the first and second derivatives of the position give the velocity and acceleration of the particle.

```
> s:=10*t-20*t^2+2*t^3;
```
$$s := 10\,t - 20\,t^2 + 2\,t^3$$

```
> v:=diff(s, t);
```
$$v := 10 - 40\,t + 6\,t^2$$

```
> a:=diff(s, t$2);
```
$$a := -40 + 12\,t$$

Symbolic integration will be performed with the *int* command. The general format of this command is int(f, x = a..b) where f is the integrand (a Maple expression), x is the integration variable and a and b are the limits of integration. If the integration limits are omitted, the indefinite integral will be evaluated. Here are several examples of definite and indefinite integrals.

```
> restart;
> int(sin(b*x),x);
```
$$-\frac{\cos(b\,x)}{b}$$

```
> int(sin(b*x),x = a..b);
```
$$\frac{-\cos(b^2) + \cos(b\,a)}{b}$$

```
> g:=b*log(x);
```
$$g := b \ln(x)$$

```
> int(g,x);
```
$$b\,x \ln(x) - b\,x$$

```
> int(g,x = c..d);
```
$$b\,d \ln(d) - b\,d - b\,c \ln(c) + b\,c$$

If a definite integral contains no unknown parameters either in the integrand or the integration limits, the *int* command will provide numerical answers. Here are a few examples.

> int(x+3*x^3, x = 0..3);
$$\frac{261}{4}$$

> int(log(x), x = 2..5); # don't forget that log is the natural logarithm
$$-3 + 5 \ln(5) - 2 \ln(2)$$

Note that Maple will always try to return an exact answer. This usually results in answers containing fractions or functions as in the above examples. This is very useful in some situations; however, one often wants to know the numerical answer without having to evaluate a result such as the above with a calculator. To obtain numerical answers use Maple's *evalf* function as in the following examples.

> evalf(int(x+3*x^3, x = 0..3));
$$65.25000000$$

> evalf(int(log(x), x = 2..5));
$$3.660895199$$

## 1.6 Solving Equations

Solving one or more equations in Maple is usually accomplished with the solve command. The calling sequence for the solve command is solve(eqns, vars) where "eqns" is a set of equations and "vars" is a set of variables to solve for. There may be more unknowns in a set of equations than appear in the "vars" list. If the number of equations equals the number of unknowns, Maple will return a numerical answer. If the number of equations is less than the number of unknowns, Maple will evaluate the expression symbolically.

### Solving Single Equations

Let's consider the simple case of finding the roots of a quadratic equation, $ax^2 + bx + c = 0$.

> restart;
> solve(a*x^2+b*x+c = 0,x);
$$-\frac{b - \sqrt{b^2 - 4ac}}{2a}, \ -\frac{b + \sqrt{b^2 - 4ac}}{2a}$$

Here is another example.

> v:=3-6*t+2*t^2;
$$v := 3 - 6t + 2t^2$$

We have a general expression for v as a function of t. We can easily find the value of t at which v = 2 as follows.

```
> solve(v=2,t);
```
$$\frac{3}{2}+\frac{1}{2}\sqrt{7}, \frac{3}{2}-\frac{1}{2}\sqrt{7}$$

If we don't want to get our calculators out to evaluate the above expressions for t, we should use the following:

```
> evalf(solve(v=2,t));
```
$$2.822875656 \ .177124344$$

We can also use the solve command to get a general expression for t given any specified v = v0.

```
> solve(v=v0,t);
```
$$\frac{3}{2}+\frac{1}{2}\sqrt{3+2\,v0}, \frac{3}{2}-\frac{1}{2}\sqrt{3+2\,v0}$$

As another example, assume v is a function of an angle $\theta$ and we would like to find the values of $\theta$ at which v has a specific value:

```
> v:=4*sin(2*theta)-2*cos(2*theta);
```
$$v := 4\sin(2\,\theta) - 2\cos(2\,\theta)$$

```
> solve(v=v0,theta);
```
$$\frac{1}{2}\arctan\left(\frac{v0}{5}+\frac{\sqrt{-v0^2+20}}{10}, -\frac{v0}{10}+\frac{\sqrt{-v0^2+20}}{5}\right),$$
$$\frac{1}{2}\arctan\left(\frac{v0}{5}-\frac{\sqrt{-v0^2+20}}{10}, -\frac{v0}{10}-\frac{\sqrt{-v0^2+20}}{5}\right)$$

Occasionally when solving equations in Maple you may encounter the following.

```
> restart;
> f:=10*sin(x)=2*x^2+1;
```
$$f := 10\sin(x) = 2\,x^2 + 1$$

```
> solve(f,x);
```
$$\text{RootOf}(-10\sin(\_Z)+2\,\_Z^2+1, label=\_L1)$$

It is beyond the scope of this chapter to get into exactly what the *RootOf* result means in Maple. Suffice it to say that Maple was not able to find an exact (symbolic) solution of the equation. There are two general approaches to obtaining an approximate solution that you might consider in a case like this; graphical and numerical.

With a graphical approach you could consider making a plot of the expressions on either side of the equals sign and then finding where the two curves intersect.

> plot([10*sin(x), 2*x^2+1], x=0...3);

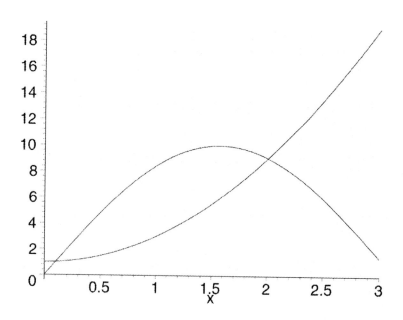

From the plot above we see that there are two solutions at x equals about 0.1 and 2.0. Of course, you might have to experiment around with the plot limits to make sure that you have identified all the solutions. To obtain a more accurate answer you should print the graph and then draw vertical lines at the intersections of the two curves. The intersections of these lines with the x-axis will give the two solutions. Another approach, which avoids having to draw vertical lines, is to rewrite the equation so that you have zero on the right hand side.

> g:=10*sin(x)-2*x^2-1;
$$g := 10 \sin(x) - 2 x^2 - 1$$

Observe that the values of x for which g = 0 will be solutions to our original equation. Now plot g versus x and find where the curve intersects the x axis.

> plot(g, x=0..2.5);

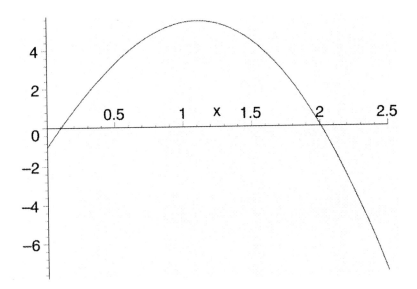

A much faster approach is to use fsolve to obtain a numerical solution.

> fsolve(f,x);
                .1022700143

This was certainly quick, but note also a danger of missing multiple solutions. Thus it is generally a good idea to make a plot even when you are going to use *fsolve*. From the plot we can pick up the second solution by using fsolve and specifying a range.

> fsolve(f,x,x=1..3);
                2.007613807

### Finding Maxima and Minima of Functions

The usual method for finding maxima or minima of a function f(x) is to first determine the location(s) x at which maxima or minima occur by solving the equation $\frac{df}{dx} = 0$ for x. One then substitutes the value(s) of x thus determined into f(x) to find the maximum or minimum. Consider finding the maximum velocity given the following expression for the velocity as a function of time:

> v:=b+c*t-d*t^3;
                $v := b + c\,t - d\,t^3$

> tm:=solve(diff(v,t)=0,t);
                $tm := \dfrac{\sqrt{3}\,\sqrt{d\,c}}{3\,d}, \; -\dfrac{\sqrt{3}\,\sqrt{d\,c}}{3\,d}$

> vm:=subs(t=tm[1],v);

$$vm := b + \frac{c\sqrt{3}\sqrt{dc}}{3d} - \frac{\sqrt{3}(dc)^{(3/2)}}{9d^2}$$

Note that t = tm[1] substitutes the first of the two results for tm. The second root is negative and thus not physically significant. Whether the result vm corresponds to a minimum or a maximum depends on the values of b, c and d. Let's look at a specific case:

> b:=1: c:=2: d:=1:
> tm[1];vm;

$$\frac{1}{3}\sqrt{3}\sqrt{2}$$

$$1 + \frac{4}{9}\sqrt{3}\sqrt{2}$$

In the present case, it is fairly clear that $v = 1 + \frac{4\sqrt{6}}{9} = 2.0887$ is a maximum. It never hurts to check, though. An easy way to do this is to substitute neighboring values of t:

> evalf(subs(t=tm[1]-0.01,v));evalf(subs(t=tm[1]+0.01,v));
                 2.088418159

                 2.088416159

Both values are somewhat less than 2.0887, verifying that we have found a maximum. Probably the safest way to verify a maximum, though, is to plot the results.

> plot(v,t=0..1.8, labels=[`time`,`velocity`]);

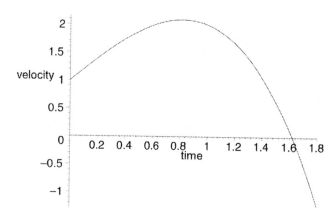

A simpler way to find maxima and minima is to use the exrema function.

> extrema(v,{},t);
$$\{1 - \frac{4}{9}\sqrt{6}, 1 + \frac{4}{9}\sqrt{6}\}$$

Note that Maple has found both a maximum and a minimum. Further investigation would show that the minimum occurs at the negative time root found above. The brackets {} in the extrema function allow the input of constraints. The problems that we will consider will not have constraints, thus the brackets will be left empty.

### Systems of Equations

> restart;
> eqn1:=-P*sin(beta)+Bx-Ax=0;
  eqn2:=Ay+P*cos(beta)-w*a=0;
  eqn3:=P*a*cos(beta)-P*b*sin(beta)+Bx*c-1/2*w*a^2=0;

$$eqn1 := -P\sin(\beta) + Bx - Ax = 0$$

$$eqn2 := Ay + P\cos(\beta) - wa = 0$$

$$eqn3 := Pa\cos(\beta) - Pb\sin(\beta) + Bxc - \frac{1}{2}wa^2 = 0$$

Above we have three equations that we are going to solve for three variables using Maple's solve command. Since we have more unknowns than we have equations, Maple will solve the equations symbolically in terms of the unknown parameters. We can ask Maple to solve for any three of the unknowns we wish. Here we will ask Maple to solve for Ax, Ay, and Bx and will obtain three expressions in terms of the remaining parameters P, w, a, b, c and $\beta$ .

> soln:=solve({eqn1,eqn2,eqn3},{Ax,Ay,Bx});
$$soln := \{Ay = -P\cos(\beta) + wa, Bx = \frac{1}{2}\frac{-2Pa\cos(\beta) + 2Pb\sin(\beta) + wa^2}{c},$$

$$Ax = \frac{1}{2}\frac{-2P\sin(\beta)c - 2Pa\cos(\beta) + 2Pb\sin(\beta) + wa^2}{c}\}$$

It is important to understand how Maple goes about assigning numbers or symbolic expressions to names. Even though the above output has the form Ax = expression, Maple has not yet assigned anything to the name Ax. To see this we type Ax; in the next line and Maple returns simply Ax indicating that nothing is assigned to the name Ax.

> Ax;
$$Ax$$

The assign statement is used to assign the names Ax, Ay and Bx to expressions in soln:

> assign(soln);

The following statements are unnecessary, merely showing explicitly that the assignments have been made.

> Ax; Ay; Bx;

$$\frac{1}{2}\frac{-2\,P\sin(\beta)\,c - 2\,P\,a\cos(\beta) + 2\,P\,b\sin(\beta) + w\,a^2}{c}$$

$$-P\cos(\beta) + w\,a$$

$$\frac{1}{2}\frac{-2\,P\,a\cos(\beta) + 2\,P\,b\sin(\beta) + w\,a^2}{c}$$

We are now ready to do a parametric study. As one example, we will assume values for P, w, a, b and c and then plot Ax, Ay, and Bx as a function of $\beta$ .

> P:=300:w:=50:a:=10:b:=5:c:=20:
> with(plots):
> t1:=textplot([4.2,450,`A x`]):t2:=textplot([2.7,350,`B x`]):t3:=textplot([3.2,870,`A y`]):
> plot1:=plot([Ax,Ay,Bx],beta=0..2*Pi):
> display(plot1,t1,t2,t3); ·

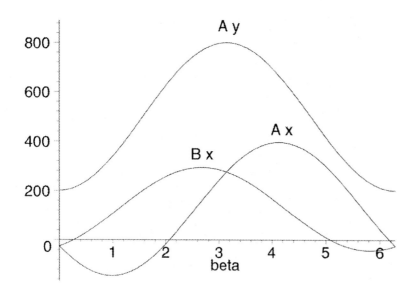

# KINEMATICS OF PARTICLES

# 2

Kinematics involves the study of the motion of bodies irrespective of the forces that may produce that motion. Maple can be very useful in solving particle kinematics problems. Problem 2.1 is a rectilinear motion problem illustrating integration with the *int* command. The formulation of this problem results in an equation that cannot be solved exactly except with some rather sophisticated mathematics. When this occurs it is generally easiest to obtain either a graphical or numerical solution. This problem illustrates both approaches. Problem 2.2 is a rectangular coordinates problem that illustrates *diff* and *solve*. Problem 2.3 is a relatively straightforward problem where Maple is used to generate a plot that might be useful in a parametric study. The path of a particle is depicted using a parametric plot and a polar plot in problem 2.4. In problem 2.5, the $r$-$\theta$ components of the velocity are determined using symbolic differentiation (*diff*). The problem also illustrates how computer algebra can simplify what might normally be a rather tedious algebra problem. The *diff* command is further illustrated in problems 2.6 and 2.7. Problem 2.7 is particularly interesting in that it requires differentiation with respect to time of a function whose explicit time dependence is unknown. This happens rather frequently in Dynamics so it is useful to know how to accomplish this with Maple.

## 2.1 Sample Problem 2/4 (Rectilinear Motion)

A freighter is moving at a speed of 8 knots when its engines are suddenly stopped. From this time forward, the deceleration of the ship is proportional to the square of its speed, so that $a = -kv^2$. The sample problem in your text shows that it is rather easy to determine the constant $k$ by measuring the speed of the boat at some specified time. Show how $k$ could be found by (a) measuring the speed after some specified distance and (b) measuring the time required to travel some specified distance. In both cases let the initial speed be $v_0$.

### Problem Formulation

(a) Since time is not involved, the easiest approach is to integrate the equation $v dv = a ds$.

$$v dv = a ds = -kv^2 ds \qquad \int_{v_0}^{v} \frac{dv}{v} = -k \int_{0}^{s} ds \qquad ks = \ln\left(\frac{v_0}{v}\right)$$

With this result it is easy to find $k$ given $v$ at some specified $s$. To illustrate, assume that $v_0 = 8$ knots and that the speed of the boat is determined to be 3.9 knots after it has traveled one nautical mile.

$$k(1) = \ln\left(\frac{8}{3.9}\right) \qquad k = 0.718 \text{ mi}^{-1}$$

(b) Here we follow the general approach in the sample problem. Integrating $a = dv/dt$ yields

$$\int_{v_0}^{v} \frac{dv}{v^2} = -k \int_{0}^{t} dt \qquad -kt = \frac{v - v_0}{vv_0} \qquad v = \frac{v_0}{1 + ktv_0}$$

To obtain the distance $s$ as a function of time we integrate $v = ds/dt$

$$\int_{0}^{s} ds = s = \int_{0}^{t} v dt = \int_{0}^{t} \frac{v_0}{1 + ktv_0} dt \qquad s = \frac{1}{k} \ln(1 + ktv_0)$$

This equation turns out to be very difficult to solve for $k$. A good mathematician or someone familiar with symbolic algebra software might be able to find the general solution for $k$ in terms of the so-called LambertW function (LambertW(x) is the solution of the equation $ye^y = x$). Even if this solution were found it would be of little use in most practical situations. For example, you would have to spend some time familiarizing yourself with the function. Once this is done you would still have to use a program like Maple or a mathematical handbook to evaluate the function.

For these reasons it is probably easiest to find $k$ either graphically or numerically. Obtaining a numerical solution with Maple is so easy that there is little reason not to use this approach. It is generally advisable though to use a graphical approach even when a numerical solution is being obtained. This is the best way to identify whether there are multiple solutions to the problem and also serves as a useful check on the numerical results. Thus, both approaches are illustrated in the worksheet below.

The usual way to generate a graphical solution is to rearrange the equation so as to give a function that is zero at points that are solutions to the original equation. Rearranging the equation above in this manner yields,

$$f = ks - \ln(1 + ktv_0) = 0$$

Given values of $s$, $t$, and $v_0$, $f$ can be plotted versus $k$. The value of $k$ at which $f = 0$ provides the solution to the original equation.

## Maple Worksheet

> restart;

Although the integrations are simple in this problem, we'll go ahead and evaluate them symbolically for purposes of illustration.

> s_a:=-1/k*int(1/x,x=v0..v);
Warning, unable to determine if 0 is between v0 and v; try to use assumptions or set _EnvAllSolutions to true

$$s\_a := -\left( \frac{1}{k} \int_{v0}^{v} \frac{1}{x} \, dx \right)$$

Note that Maple did not evaluate the integral symbolically. In some cases Maple gets a little irritating in that it cannot actually "understand" things about the problem that are to us trivially obvious. In the present case (read the warning message) Maple is not sure whether 0 is between v0 and v. One way around this is to use **assume** to tell Maple that v and v0 are both positive. While we are at it we will also assume k and t are positive so we won't encounter problems with the second integral.

```
> assume(v0>0);assume(v>0);assume(k>0);assume(t>0);
> s_a:=-1/k*int(1/x,x=v0..v);
```

$$s\_a := -\frac{-\ln(v0\!\sim) + \ln(v\!\sim)}{k\!\sim}$$

```
> s_b:=v0*int(1/(1+k*v0*x),x=0..t);
```

$$s\_b := \frac{\ln(1 + t\!\sim k\!\sim v0\!\sim)}{k\!\sim}$$

The tilde ($\sim$) follows any variable for which something has been assumed.

It should also be pointed out that this problem only occurs in later editions of Maple. For example, Maple 6 evaluates the two integrals without needing any *assume* statements. Now we move on to illustrate a graphical and numerical solution to our problem.

To illustrate the graphical solution, take v0 = 8 knots and assume that the boat is found to move 1.1 nautical miles after 10 minutes.

```
>
> v0:=8: s:=1.1: t:=10/60:
> f:=k*s-log(1+k*v0*t);
```

$$f := 1.1\, k\!\sim\, -\ln\!\left(1 + \frac{4\, k\!\sim}{3}\right)$$

```
> plot(f,k=0..0.5,color=black);
```

> plot(f, k = 0..0.5);

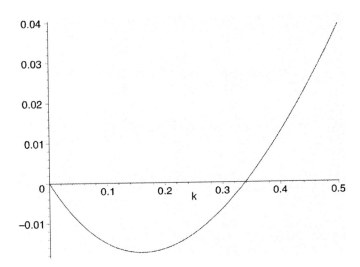

The above graph shows that $k$ is about 0.34 mi$^{-1}$. Now let's try a numerical solution with *fsolve*.

> fsolve(f = 0,k);
       0.

Here we have found the solution $k=0$ which is not physically significant. Let's try again specifying a range.

> fsolve(f = 0, k, 0.3..0.4);
       .3392305334

This illustrates the importance of the graphical approach. If we had not first plotted f versus k we would have a hard time specifying a range for Maple to search for a root. Now let's see what we get using the symbolic *solve* command.

> solve(f=0,k);
       0., .3392305334

Since Maple was unable to find an exact (symbolic) solution to the equation it automatically reverted to a numerical approach and actually found both solutions. We showed both approaches (*solve* and *fsolve*) in this problem since it doesn't always turn out this nicely.

## 2.2 Problem 2/87 (Rectangular Coordinates)

A long-range rifle is fired at $A$ with the projectile hitting the mountain at $B$. (a) If the muzzle velocity is $u$ = 400 m/s, determine the two angles of elevation $\theta$ which will permit the projectile to hit the mountain target $B$ and plot the two trajectories. (b) Determine the smallest muzzle velocity that will allow the projectile to strike at $B$ and the angle at which it must be fired. Repeat the plot for part (a) and include the trajectory of the projectile for this minimum initial velocity.

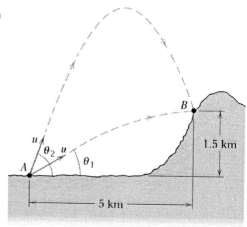

### Problem Formulation

Place a coordinate system at $A$ with $x$ positive to the right and $y$ positive up. The initial components of the velocity are,

$$(v_x)_0 = u\cos\theta \qquad (v_y)_0 = u\sin\theta$$

The acceleration is constant with components $a_x = 0$ and $a_y = -g$. Integrating these two accelerations twice and applying initial conditions yields (see page 44 of your text if you need additional details),

$$x = u\cos\theta\, t \qquad\qquad y = u\sin\theta\, t - 1/2gt^2$$

Plotting $y$ in terms of $x$ for different times $t$ will yield the trajectory of the projectile. This type of plot is called a **parametric plot** since the items plotted ($x$ and $y$) are each known in terms of another parameter ($t$).

Anytime you have a projectile motion problem and you know the coordinates of a point on the trajectory (our point $B$) you should solve for $x$ and $y$ (as we have done above) and then obtain two equations by substituting the coordinates of the points. These two equations can then be solved for two unknowns. Note that in most cases one of the two unknowns will be the time of flight.

*Part (a)* Substituting $x$ = 5,000 m, $y$ = 1,500 m and $u$ = 400 m/s gives

$$5000 = 400\cos\theta\, t \qquad 1500 = 400\sin\theta\, t - 1/2(9.81)t^2$$

We will let Maple solve these two equations simultaneously. Maple actually finds four solutions, however two of the four can be discarded since they involve negative times. The other two solutions correspond to the two solutions shown in the illustration accompanying the problem statement. The results are,

$$\theta_1 = 26.6° \text{ and } \theta_2 = 80.6°$$

*Part (b)* It should be intuitively obvious why there must be a minimum initial velocity below which the projectile cannot reach *B*. How do we go about finding it? We still have the two equations for the coordinates of point *B*,

$$5000 = u\cos\theta\, t \qquad 1500 = u\sin\theta\, t - 1/2(9.81)t^2$$

however there are now three unknowns ($u$, $\theta$, $t$). Suppose for the moment that the launch angle $\theta$ were given and we were asked to calculate the required initial speed $u$ so that the projectile strikes *B*. In this case we would have two equations and two unknowns. From this observation we see that $u$ is a function of $\theta$ from which we get our general solution strategy:

(a)     Eliminate $t$ from the above two equations and solve for $u$ as a function of $\theta$.

(b)     Differentiate this function with respect to $\theta$ to find the location of the minimum.

Solving the first equation for u gives

$$u = \frac{5000}{t\cos\theta}$$

Substituting into the second yields $1500 = 5000\tan\theta - \frac{1}{2}gt^2$. This equation is now solved for $t = \sqrt{2(5000\tan\theta - 1500)/g}$ which can be substituted back into *u* to give

$$u = \frac{5000}{\cos\theta\sqrt{2(5000\tan\theta - 1500)/g}}$$

We will let Maple differentiate this equation and solve for the minimum speed and the associated launch angle. The result is

$$u_{min} = 256.8 \text{ m/s at } \theta = 53.3°$$

## *Maple Worksheet*

```
> restart; with(plots):
> x:=u*cos(theta)*t;
```
$$x := u \cos(\theta)\, t$$

```
> y:=u*sin(theta)*t-1/2*g*t^2;
```
$$y := u \sin(\theta)\, t - \frac{g\, t^2}{2}$$

```
> u:=400: g:=9.81:
> solve([x=5000,y=1500],[theta,t]);
```
$$[[\theta = 0.4556954358,\ t = 13.92051641\,],\ [\theta = 1.406557685,\ t = 76.45200773\,],$$
$$[\theta = -2.685897218,\ t = -13.92051641\,],\ [\theta = -1.735034968,\ t = -76.45200773\,]]$$

Note that Maple has found four solutions but that two can be excluded because they have time being negative. The two angles are substituted back into *x* and *y* to get the two trajectories.

```
> x1:=subs(theta=0.455695,x);y1:=subs(theta=0.455695,y);
```
$$x1 := 400 \cos(0.455695)\, t$$

$$y1 := 400 \sin(0.455695)\, t - 4.905000000\, t^2$$

```
> x2:=subs(theta=1.4066,x);y2:=subs(theta=1.4066,y);
```
$$x2 := 400 \cos(1.4066)\, t$$

$$y2 := 400 \sin(1.4066)\, t - 4.905000000\, t^2$$

```
> p1:=plot([x1/1000,y1/1000,t=0..13.92],0..6,0..8,labels=["x (km)","y (km)"]):
> p2:=plot([x2/1000,y2/1000,t=0..76.45],0..6,0..8):
```

Note the special format required for the parametric plot. In particular, note how each parametric plot uses a different time parameter (13.92 and 76.45). This is necessary to ensure that the plots stop at point *B*.

> display(p1,p2);

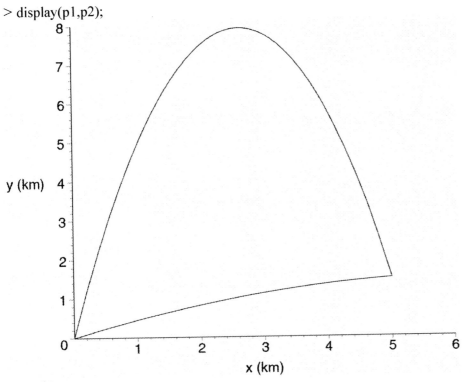

part (b)

> unassign('u');
> t[B]:=sqrt(2*(5000*tan(theta)-1500)/g);

$$t_B := \sqrt{1019.367992\,\tan(\theta) - 305.8103976}$$

> u:=5000/cos(theta)/t[B];

$$u := \frac{5000}{\cos(\theta)\sqrt{1019.367992\,\tan(\theta) - 305.8103976}}$$

> du:=diff(u,theta);

$$du := \frac{5000\,\sin(\theta)}{\cos(\theta)^2\sqrt{1019.367992\,\tan(\theta) - 305.8103976}}$$
$$- \frac{2500\,(1019.367992 + 1019.367992\,\tan(\theta)^2)}{\cos(\theta)\,(1019.367992\,\tan(\theta) - 305.8103976)^{(3/2)}}$$

> solve(du=0,theta);

$$0.9311265606,\ -0.6396697662,\ -2.210466093,\ 2.501922887$$

Only one of the four solutions is between 0 and 90°.

> evalf(.9311*180/Pi); # elevation angle
                53.34810029

> u:=evalf(subs(theta=.9311,u)); # minimum muzzle velocity
                $u := 256.7580644$

> evalf(subs(theta=.9311,t[B])); # time to reach B
                32.62170207

> x3:=subs(theta=.9311265606,x):y3:=subs(theta=.9311265606,y):
> p3:=plot([x3/1000,y3/1000,t=0..32.62],0..6,0..8,color=black):
> display(p1,p2,p3);

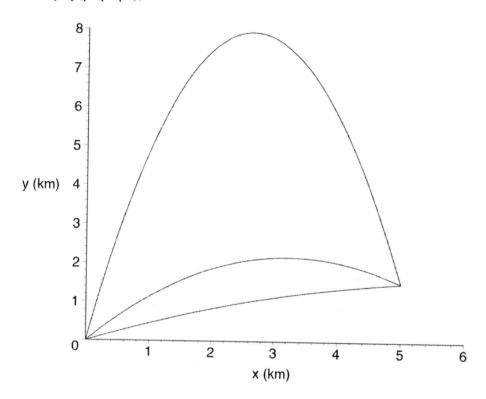

## 2.3 Problem 2/120 (n-t Coordinates)

A baseball player releases a ball with initial conditions shown in the figure. Plot the radius of curvature of the path just after release and at the apex as a function of the release angle $\theta$. Explain the trends in both results as $\theta$ approaches 90°.

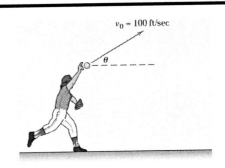

### Problem Formulation

*Just after release*

$$a_n = g\cos\theta = \frac{v_0^2}{\rho} \qquad \rho = \frac{v_0^2}{g\cos\theta}$$

*At the apex*

At the apex, $v_y = 0$ and $v = v_x = v_0\cos\theta$. Since $v$ is horizontal, the normal direction is vertically downward so that $a_n = g$.

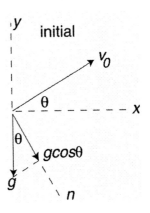

$$a_n = g = \frac{(v_0\cos\theta)^2}{\rho} \qquad \rho = \frac{(v_0\cos\theta)^2}{g}$$

### Maple Worksheet

```
> restart; with(plots):
> rho[i]:=v0^2/g/cos(theta);
```
$$\rho_i := \frac{v0^2}{g\cos(\theta)}$$

```
> vx:=v0*cos(theta0);
```
$$vx := v0\cos(\theta0)$$

```
> rho[a]:=(v0*cos(theta))^2/g;
```
$$\rho_a := \frac{v0^2\cos(\theta)^2}{g}$$

```
> v0:=100: g:=32.2:
> p:=plot([rho[i],rho[a]],theta=0..Pi/2,y=0..800,labels=["theta (rads)",""],title=
"radius of curvature (ft)"):
> t1:=textplot([1,500,"initial"]):
> t2:=textplot([1,140,"apex"]):
```

> display(p,t1,t2);

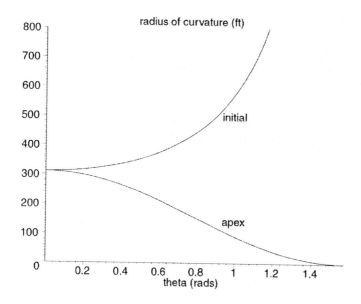

Note that as $\theta$ approaches 90° ($\pi/2$), the initial $\rho$ goes to infinity while $\rho$ at the apex approaches zero. When $\theta = 90°$, the ball travels along a straight (vertical) path. As you recall, straight paths have a radius of curvature of infinity. At the apex, the velocity will be zero giving a radius of curvature of zero.

## 2.4 Sample Problem 2/9 (Polar Coordinates)

Rotation of the radially slotted arm is governed by $\theta = 0.2t + 0.02t^3$, where $\theta$ is in radians and $t$ is in seconds. Simultaneously, the power screw in the arm engages the slider $B$ and controls its distance from $O$ according to $r = 0.2 + 0.04t^2$, where $r$ is in meters and $t$ is in seconds. Calculate the magnitudes of the velocity and acceleration of the slider as a function of time $t$. (a) Plot $v$, $v_r$ and $v_\theta$ for $t$ between 0 and 5 sec. (b) Plot $a$, $a_r$ and $a_\theta$ for $t$ between 0 and 5 sec. (c) Plot the path of the slider $B$ and compare with the result in your book.

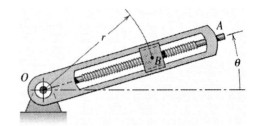

### Problem Formulation

The first part of this problem solution will be identical to that in the Sample Problem in your text except that everything will be left in terms of $t$. To summarize,

$$r = 0.2 + 0.04t^2 \qquad \dot{r} = 0.08t \qquad \ddot{r} = 0.08$$

$$\theta = 0.2t + 0.02t^3 \qquad \dot{\theta} = 0.2 + 0.06t^2 \qquad \ddot{\theta} = 0.12t$$

Now all we have to do is substitute these expressions into the definitions for the velocity and acceleration. As usual, there is no need to make an explicit substitution when using the computer.

$$v_r = \dot{r} = 0.08t \qquad v_\theta = r\dot{\theta} \qquad v = \sqrt{v_r^2 + v_\theta^2}$$

$$a_r = \ddot{r} - r\dot{\theta}^2 \qquad a_\theta = r\ddot{\theta} + 2\dot{r}\dot{\theta} \qquad a = \sqrt{a_r^2 + a_\theta^2}$$

The plot for part (c) can be found in two ways. The first is to use the suggestion in your book and write

$$x = r\cos\theta \qquad y = r\sin\theta$$

Now we have the $x$ and $y$ coordinates of the slider in terms of a common parameter $t$. This suggests that we can use a parametric plot. Also, the fact that we are using polar coordinates would indicate that we might use Maple's *polarplot*. This provides us with the second plotting method.

*Maple Worksheet*

```
> restart; with(plots):
> r:=0.2+0.04*t^2;
```
$$r := 0.2 + 0.04 \ t^2$$

```
> rd:=diff(r,t);
```
$$rd := 0.08 \ t$$

```
> rdd:=diff(rd,t);
```
$$rdd := 0.08$$

```
> theta:=0.2*t+0.02*t^3;
```
$$\theta := 0.2 \ t + 0.02 \ t^3$$

```
> thetad:=diff(theta,t);
```
$$thetad := 0.2 + 0.06 \ t^2$$

```
> thetadd:=diff(thetad,t);
```
$$thetadd := 0.12 \ t$$

```
> vr:=rd; vtheta:=r*thetad;
```
$$vr := 0.08 \ t$$

$$vtheta := ( \, 0.2 + 0.04 \ t^2 \, ) \, ( \, 0.2 + 0.06 \ t^2 \, )$$

```
> v:=sqrt(vr^2+vtheta^2);
```
$$v := \sqrt{ \, 0.0064 \ t^2 + ( \, 0.2 + 0.04 \ t^2 \, )^2 \, ( \, 0.2 + 0.06 \ t^2 \, )^2 \, }$$

```
> ar:=rdd-r*thetad^2;
```
$$ar := 0.08 - ( \, 0.2 + 0.04 \ t^2 \, ) \, ( \, 0.2 + 0.06 \ t^2 \, )^2$$

```
> atheta:=r*thetadd+2*rd*thetad;
```
$$atheta := 0.12 \ ( \, 0.2 + 0.04 \ t^2 \, ) \, t + 0.16 \ t \, ( \, 0.2 + 0.06 \ t^2 \, )$$

```
> a:=sqrt(ar^2+atheta^2): # output suppressed
```

*Part (a)*
```
> p1:=plot([v,vr,vtheta],t=0..5,color=black,labels=["time (s)",""],title="part (a) velocity (m/s)"):
> t1:=textplot([3.35,.75,'v']): t2:=textplot([4.1,.75,'vtheta']):
> t3:=textplot([4.1,.4,'vr']):
> display(p1,t1,t2,t3);
```

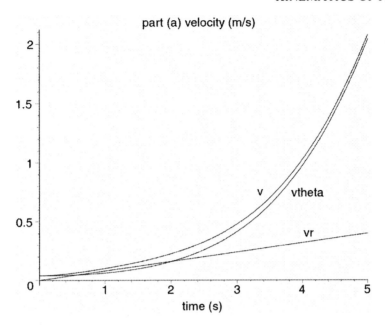

*Part (b)*
> p2:=plot([a,ar,atheta],t=0..5,color=black,labels=["time (s)",""],title="part (b) acceleration (m/s^2)"):
> t4:=textplot([4,2,'a']): t5:=textplot([4.5,-1.4,'ar']):
> t6:=textplot([4.5,1.2,'atheta']):
> display(p2,t4,t5,t6);

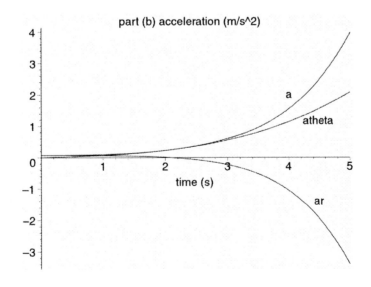

*Part (c)*
> x:=r*cos(theta);y:=r*sin(theta);

$$x := ( 0.2 + 0.04 \ t^2 ) \cos ( 0.2 \ t + 0.02 \ t^3 )$$

$$y := ( 0.2 + 0.04 \ t^2 ) \sin ( 0.2 \ t + 0.02 \ t^3 )$$

> plot([x,y,t=0..5],color=black,title="part (c) method 1: parametric plot");

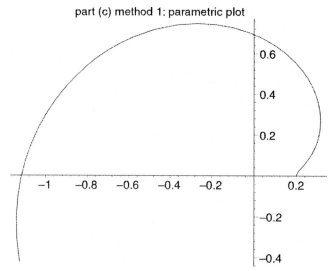

part (c) method 1: parametric plot

> polarplot([r,theta,t=0..5],color=black,title="part (c) method 2: polar plot");

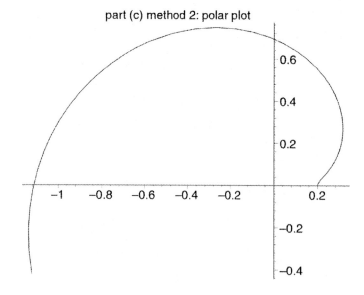

part (c) method 2: polar plot

## 2.5 Sample Problem 2/10 (Polar Coordinates)

A tracking radar lies in the vertical plane of the path of a rocket which is coasting in unpowered flight above the atmosphere. For the instant when $\theta = 30°$, the tracking data give $r = 25(10^4)$ feet, $\dot{r} = 4000$ ft/s, and $\dot{\theta} = 0.8$ deg/s. Let this instant define the initial conditions at time t = 0 and plot $v_r$ and $v_\theta$ as a function of time for the next 150 seconds. You may assume that g remains constant at 31.4 ft/s$^2$ during this time interval.

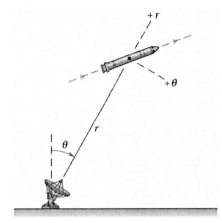

### Problem Formulation

Place a Cartesian coordinate system at the radar with $x$ positive to the right and $y$ positive up. Since the rocket is coasting in unpowered flight we can use the equations for projectile motion

$$x = x_0 + v_0 \cos(\beta)t \qquad y = y_0 + v_0 \sin(\beta)t - \frac{1}{2}gt^2$$

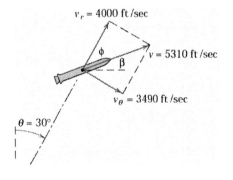

Where $x_0 = 25(10^4)\sin(30)$ ft, $y_0 = 25(10^4)\cos(30)$ ft, $v_0$ is the initial speed (5310 ft/sec, see the sample problem) and $\beta$ is the angle that $v_0$ makes with the horizontal. From the figure shown to the right we can find the angle between $v_0$ and the $r$ axis as $\phi = \tan^{-1}(3490/4000) = 41.11°$. Since the $r$ axis is 60° from the horizontal, $\beta = 60 - 41.11 = 18.89°$.

With $r$ and $\theta$ defined as in the sample problem we have, at any time t

$$r = \sqrt{x^2 + y^2} \qquad \theta = \tan^{-1}(x/y)$$

Now we find $v_r$ and $v_\theta$ from their definitions.

$$v_r = \dot{r} \qquad\qquad v_\theta = r\dot{\theta}$$

Substitution of $x$ and $y$ into the above equations and carrying out the derivatives with respect to time gives $v_r$ and $v_\theta$ as functions of time. The results are very messy and will not be given here. Remember, though, that substitutions such as this can be made automatically when using computer software such as Maple.

### Maple Worksheet

```
> restart; with(plots):
> x:=x0+v0*cos(beta)*t;
```
$$x := x0 + v0 \cos(\beta)\, t$$

```
> y:=y0+v0*sin(beta)*t-1/2*g*t^2;
```
$$y := y0 + v0 \sin(\beta)\, t - \frac{1}{2} g\, t^2$$

```
> r:=sqrt(x^2+y^2);
```
$$r := \sqrt{(x0 + v0 \cos(\beta)\, t)^2 + \left( y0 + v0 \sin(\beta)\, t - \frac{1}{2} g\, t^2 \right)^2}$$

```
> theta:=arctan(x/y);
```
$$\theta := \arctan\left( \frac{x0 + v0 \cos(\beta)\, t}{y0 + v0 \sin(\beta)\, t - \frac{1}{2} g\, t^2} \right)$$

```
> vr:=diff(r,t);
```
$$vr := \frac{1}{2} \frac{2\,(x0 + v0 \cos(\beta)\, t)\, v0 \cos(\beta) + 2\left( y0 + v0 \sin(\beta)\, t - \frac{1}{2} g\, t^2 \right)(v0 \sin(\beta) - g\, t)}{\sqrt{(x0 + v0 \cos(\beta)\, t)^2 + \left( y0 + v0 \sin(\beta)\, t - \frac{1}{2} g\, t^2 \right)^2}}$$

```
> vtheta:=r*diff(theta,t);
```
$$vtheta := \sqrt{(x0 + v0 \cos(\beta)\, t)^2 + \left( y0 + v0 \sin(\beta)\, t - \frac{1}{2} g\, t^2 \right)^2}$$
$$\left( \frac{v0 \cos(\beta)}{y0 + v0 \sin(\beta)\, t - \frac{1}{2} g\, t^2} - \frac{(x0 + v0 \cos(\beta)\, t)(v0 \sin(\beta) - g\, t)}{\left( y0 + v0 \sin(\beta)\, t - \frac{1}{2} g\, t^2 \right)^2} \right) \Bigg/ \Bigg($$
$$1 + \frac{(x0 + v0 \cos(\beta)\, t)^2}{\left( y0 + v0 \sin(\beta)\, t - \frac{1}{2} g\, t^2 \right)^2} \Bigg)$$

```
> v0:=5310: beta:=18.89*Pi/180: x0:=125000: y0:=216506: g:=31.4:
> p1:=plot([vr,vtheta],t=0..150, labels=[`t (sec)`,`v (ft/s)`],
labeldirections=[HORIZONTAL,VERTICAL]):
> t1:=textplot([30,4350,`v`],font=[TIMES,ITALIC,13]):
> t2:=textplot([33,4300,'r'],font=[TIMES,ITALIC,13]):
> t3:=textplot([30,3000,`v`],font=[TIMES,ITALIC,13]):
```

> t4:=textplot([33,2930,`q`],font=[SYMBOL,11]):
>
> display(p1,t1,t2,t3,t4);

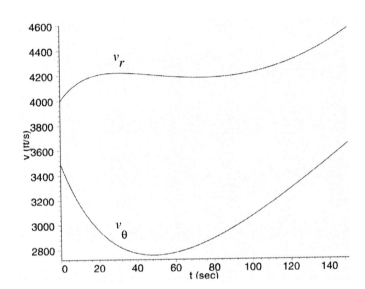

## 2.6 Problem 2/183 (Space Curvilinear Motion)

The base structure of the firetruck ladder rotates about a vertical axis through $O$ with a constant angular velocity $\dot\theta = \Omega$. At the same time, the ladder unit $OB$ elevates at a constant rate $\dot\phi = \Psi$, and section $AB$ of the ladder extends from within section $OA$ at the constant rate $\dot R = \Lambda$. Find general expressions for the components of acceleration of point $B$ in spherical coordinates if, at time $t = 0$, $\theta = 0$, $\phi = 0$, and $AB = 0$. Express your answers in terms of $\Omega$, $\Psi$, $\Lambda$, $R_0$ and t, where $R_0 = OA$ and is constant. Plot the components of acceleration of $B$ as a function of time for the case $\Omega=10$ deg/s, $\Psi= 7$ deg/s, $\Lambda = 0.5$ m/s, and $R_0 = 9$ m. Let t vary between 0 and the time at which $\phi = 90°$.

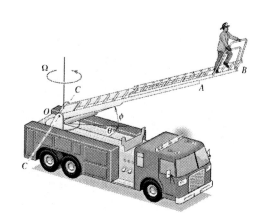

### Problem Formulation

The components of acceleration in spherical coordinates are,

$$a_R = \ddot R - R\dot\phi^2 - R\dot\theta^2 \cos^2\phi$$

$$a_\theta = \frac{\cos\phi}{R}\frac{d}{dt}\left(R^2\dot\theta\right) - 2R\dot\theta\dot\phi\sin\phi$$

$$a_\phi = \frac{1}{R}\frac{d}{dt}\left(R^2\dot\phi\right) + R\dot\theta^2 \sin\phi\cos\phi$$

The components may be obtained as functions of time by substituting,

$$R = R_0 + \Lambda t, \quad \theta = \Omega t \text{ and } \phi = \Psi t$$

Differentiation and substitution will be performed in Maple. The results are,

$$a_R = \left(R_0 + \Lambda t\right)\left(\Psi^2 - \Omega^2 \cos^2\left(\Psi t\right)\right)$$

$$a_\theta = 2\Omega\Lambda \cos(\Psi t) - 2\Omega\Psi\left(R_0 + \Lambda t\right)\sin(\Psi t)$$

$$a_\phi = 2\Psi\Lambda + (R_0 + \Lambda t)\Omega^2 \sin(\Psi t)\cos(\Psi t)$$

### Maple Worksheet

```
> restart;with(plots):
> unprotect('Psi');theta:=Omega*t;phi:=Psi*t;R:=R0+Lambda*t;
```

$$\theta := \Omega t$$

$$\phi := \Psi t$$

$$R := R0 + \Lambda t$$

Even though there are some obvious simplifications in this case, we still write the most general expressions for the spherical components of the acceleration. In this way we can consider other types of time dependence without modifying the worksheet.

```
> a[_R]:=diff(R,t,t)-R*diff(phi,t)^2-R*diff(theta,t)^2*cos(phi)^2;
a[_theta]:=cos(phi)/R*diff(R^2*diff(theta,t),t)-2*R*diff(theta,t)*diff(phi,t)*sin(phi);
a[_phi]:=1/R*diff(R^2*diff(phi,t),t)+R*diff(theta,t)^2*sin(phi)*cos(phi);
```

$$a_R := -(R0 + \Lambda t)\Psi^2 - (R0 + \Lambda t)\Omega^2\cos(\Psi t)^2$$

$$a_\theta := 2\cos(\Psi t)\Omega\Lambda - 2(R0 + \Lambda t)\Omega\Psi\sin(\Psi t)$$

$$a_\phi := 2\Psi\Lambda + (R0 + \Lambda t)\Omega^2\sin(\Psi t)\cos(\Psi t)$$

```
> Lambda:=0.5:Omega:=10*Pi/180:Psi:=7*Pi/180:R0:=9:
> tf:=Pi/2/Psi; # time at which φ = π/2 .
```

$$tf := \frac{90}{7}$$

```
> p1:=plot([a[_R],a[_theta],a[_phi]],t=0..tf,labels=["time (s)",""],
title="acceleration (m/s^2)"):
> t1:=textplot([11,-.5,'q'],font=[SYMBOL,13]):
t2:=textplot([11,.29,'f'],font=[SYMBOL,13]):
t3:=textplot([11,-.19,'R'],font=[TIMES,ROMAN,13]):
> display(p1,t1,t2,t3);
```

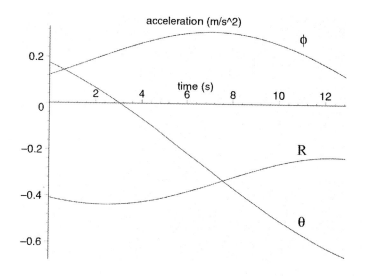

# 2.7 Sample Problem 2/16 (Constrained Motion of Connected Particles)

The tractor $A$ is used to hoist the bale $B$ with the pulley arrangement shown. If $A$ has a forward velocity $v_A$, determine an expression for the upward velocity $v_B$ of the bale in terms of $x$. Put the result in nondimensional form by introducing the velocity ratio $\eta = v_B/v_A$ and nondimensional position $\chi = x/h$. Plot $\eta$ versus $\chi$ for $0 \le \chi \le 2$.

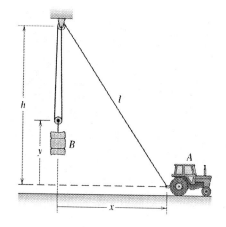

### Problem Formulation

The length $L$ of the cable can be written

$$L = 2(h - y) + l + cons \tan ts = 2(h - y) + \sqrt{h^2 + x^2} + cons \tan ts$$

Now, $\dot{L} = 0$ will be used to obtain a relation between $v_A \, (= \dot{x})$ and $v_B \, (= \dot{y})$.

$$\dot{L} = 0 = -2\dot{y} + \frac{x\dot{x}}{\sqrt{h^2 + x^2}} \qquad v_B = \frac{1}{2}\frac{xv_A}{\sqrt{h^2 + x^2}}$$

The non-dimensional result is now obtained by substituting $v_B = \eta v_A$ and $x = \chi h$.

$$\eta = \frac{1}{2} \frac{\chi}{\sqrt{1 + \chi^2}}$$

Even though these operations are rather easily performed by hand, it is instructive to have Maple do them. In particular, it will be instructive to see how to evaluate $\dot{L}$ even though $x$ and $y$ are not known explicitly as functions of time.

### Maple Worksheet

```
> restart;
> L:=2*(h-y(t))+sqrt(h^2+x(t)^2);
```

$$L := 2h - 2y(t) + \sqrt{h^2 + x(t)^2}$$

Note that we need to differentiate L with respect to time. Both x and y depend on time, however, exactly how they depend on time is not known. It turns out that this is not a problem. All we need to do is let Maple know x and y depend on time by writing x(t) and y(t).

```
> eqn:=diff(L,t);
```

$$eqn := -2\left(\frac{d}{dt}y(t)\right) + \frac{x(t)\left(\frac{d}{dt}x(t)\right)}{\sqrt{h^2 + x(t)^2}}$$

```
> vB:=solve(eqn=0,diff(y(t),t)); # solving for ydot=vB
```

$$vB := \frac{1}{2} \frac{x(t)\left(\frac{d}{dt}x(t)\right)}{\sqrt{h^2 + x(t)^2}}$$

Now we substitute vA for xdot and h $\chi$ for x.

```
> vB:=subs({diff(x(t),t)=vA,x(t)=h*chi},vB);
```

$$vB := \frac{1}{2} \frac{h\,\chi\,vA}{\sqrt{h^2 + h^2 \chi^2}}$$

Noting that h cancels, dividing through by vA (vB/vA = $\eta$ ) yields:

```
> eta:=1/2*chi/sqrt(1+chi^2);
```

$$\eta := \frac{1}{2} \frac{\chi}{\sqrt{1 + \chi^2}}$$

> plot(eta,chi=0..2,color=black,labels=["x/h",""],title="vB/vA");

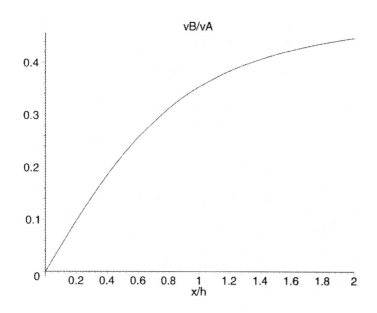

# KINETICS OF PARTICLES

# 3

The kinetics of particles is concerned with the motion produced by unbalanced forces acting on a particle. This chapter considers three approaches to the solution of particle kinetics problems: (1) direct application of Newton's second law, (2) work and energy, and (3) impulse and momentum. Problem 3.1 is a rectilinear motion problem where *solve* is used solve three equations symbolically for three unknowns. In problem 3.2, *dsolve* is used to solve a second order differential equation with initial conditions. The absolute path of a particle is then plotted using *polarplot*. Problem 3.3 uses Maple to study the effect of initial spring stretch upon the velocity of a slider. A physical interpretation of the results is also required. Problem 3.4 is a typical ballistic pendulum problem requiring both work/energy and conservation of momentum to relate the velocity of a projectile to the maximum swing angle of a pendulum. Problem 3.5 is a relatively straightforward conservation of angular momentum problem where Maple is used to generate a plot that might be useful in a parametric study. In problem 3.6, two equations are solved symbolically for two unknowns using *solve*. The maximum value of a function is then determined using *diff* and *solve*.

## 3.1 Sample Problem 3/3 (Rectilinear Motion)

The 250-lb concrete block $A$ is released from rest in the position shown and pulls the 400-lb log up the 30° ramp. Plot the velocity of the block as it hits the ground at $B$ as a function of the coefficient of kinetic friction $\mu_k$ between the log and the ramp. Let $\mu_k$ vary between 0 and 1. Why does the computer not plot results for the entire range specified?

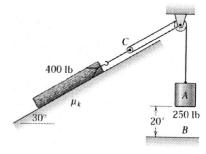

### Problem Formulation

The constant length of the cable is $L = 2s_C + s_A$ (see figure). Differentiating this expression twice yields a relation between the acceleration of $A$ and $C$ (note that $a_C = a_{LOG}$).

$$0 = 2a_C + a_A \qquad (1)$$

From the free-body diagram for the log

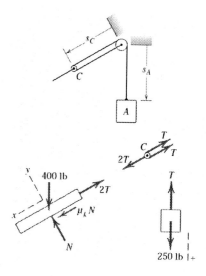

$$[\Sigma F_y = ma_y = 0] \quad N - 400\cos(30) = 0$$

$$[\Sigma F_x = ma_x] \quad \mu_k N - 2T + 400\sin(30) = \frac{400}{32.2}a_C$$

Substituting N yields,

$$400\mu_k \cos(30) - 2T + 400\sin(30) = \frac{400}{32.2}a_C \qquad (2)$$

From the free-body diagram for block $A$

$$\left[\downarrow \Sigma F = ma\right] \qquad 250 - T = \frac{250}{32.2}a_A \qquad (3)$$

Maple will be used to solve the three equations above for $a_A$, $a_C$ and $T$ in terms of $\mu_k$. Since the accelerations are constant, $v_A^2 = 2a_A d$ where $d$ is the vertical distance through which block A has fallen. Thus, the velocity of $A$ when it strikes the ground ($d = 20$ ft) is

$$v_{Af} = \sqrt{40 a_A}$$

Maple will automatically substitute $a_A$ yielding the required expression relating $v_{Af}$ and $\mu_k$.

### Maple Worksheet

> restart; Digits:=5:

> eqn1:= 0=2*aC+aA;
$$eqn1 := 0 = 2\,aC + aA$$

> eqn2:= 400*mu[k]*cos(theta)-2*T+400*sin(theta)=400/32.2*aC;
$$eqn2 := 400\,\mu_k \cos(\theta) - 2\,T + 400\sin(\theta) = 12.422\,aC$$

> eqn3:= 250-T=250/32.2*aA;
$$eqn3 := 250 - T = 7.7640\,aA$$

> theta:=30*Pi/180:

> soln:=solve({eqn1,eqn2,eqn3},{aA,aC,T});
$$soln := \{\, aA = -15.935\,\mu_k + 13.800,\ aC = 7.9675\,\mu_k - 6.9000,\ T = 142.86 + 123.72\,\mu_k \,\}$$

Note that the accelerations may be either positive or negative depending on the value of $\mu_k$. The largest value of $\mu_k$ for which the block will move up can thus be found by solving the equation aA=0 for $\mu_k$. This yields $\mu_k$ = 13.8/15.935 = 0.866.

> assign(soln);

> vAf:=sqrt(2*aA*20);
$$vAf := \sqrt{-637.40\,\mu_k + 552.00}$$

> plot(vAf,mu[k]=0..1, labels=["mu_k"," "], title="vB (ft/sec)");

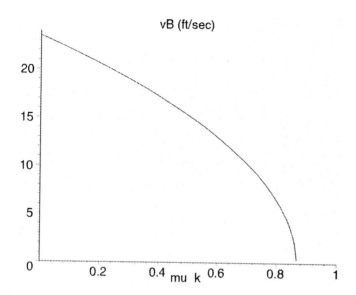

Note that results are not plotted beyond the limiting value for $\mu_k$ (0.866) that was determined above. From a numerical point of view this occurs because Maple will not plot imaginary answers. Whenever imaginary or complex values result there is usually some physical explanation. In this problem, the physical explanation is that the log will not slide up the incline if the coefficient of friction is too large.

## 3.2 Problem 3/98 (Curvilinear Motion)

The particle $P$ is released at time $t = 0$ from the position $r = r_0$ inside the smooth tube with no velocity relative to the tube, which is driven at the constant angular velocity $\omega_0$ about the vertical axis. Determine the radial velocity $v_r$, the radial position $r$, and the transverse velocity $v_\theta$ as functions of time $t$. Plot the absolute path of the particle during the time that it is inside the tube for $r_0 = 0.1$ m, $l = 1$ m, and $\omega_0 = 1$ rad/s.

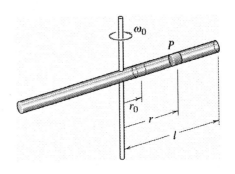

### Problem Formulation

From the free-body diagram to the right,

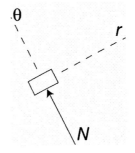

$$\Sigma F_r = 0 = ma_r = m\left(\ddot{r} - r\dot{\theta}^2\right)$$

$$\ddot{r} = r\dot{\theta}^2 = r\omega_0^2$$

Any book on differential equations will have the solution to this equation in terms of the hyperbolic sine and cosine,

$$r = A\sinh(\omega_0 t) + B\cosh(\omega_0 t)$$

The constants $A$ and $B$ are found from the initial conditions. These conditions are that $r = r_0$ and $\dot{r} = 0$ at $t = 0$. The second condition comes from the fact that the particle has no velocity (initially) relative to the tube. Before evaluating this condition we must first differentiate r with respect to time.

$$\dot{r} = A\omega_0\cosh(\omega_0 t) + B\omega_0\sinh(\omega_0 t)$$

$$r(t = 0) = r_0 = A\sinh(0) + B\cosh(0) = B$$
$$\dot{r}(t = 0) = 0 = A\omega_0\cosh(0) + B\omega_0\sinh(0) = A\omega_0$$

From the above we have $B = r_0$ and $A = 0$. Thus,

$$r = r_0\cosh(\omega_0 t)$$

From this we can obtain the radial and transverse velocities,

$$v_r = \dot{r} = r_0 \omega_0 \sinh(\omega_0 t) \qquad\qquad v_\theta = r\dot{\theta} = r_0 \omega_0 \cosh(\omega_0 t)$$

The absolute path of the particle will be graphed using polar plotting. For this we need r as a function of $\theta$. Since $\theta = \omega_0 t$ we have,

$$r = r_0 \cosh(\theta)$$

We want to plot this function only up to the point where the particle leaves the tube. Substituting $r = 1$ we have $1 = 0.1\cosh(\theta)$, or $\theta = \cosh^{-1}(10) = 2.993$ rads. Thus, the particle leaves the tube when $\theta = 2.993$ rads (171.5°).

As you will see in the worksheet below, Maple can also be used to solve the second order differential equation with initial conditions, greatly simplifying this problem.

### Maple Worksheet

> restart; with(plots):
> eqn:=diff(r(t),t,t)=omega[0]^2*r(t);

$$eqn := \frac{\partial^2}{\partial t^2} \mathrm{r}(t) = {\omega_0}^2 \, \mathrm{r}(t)$$

The following uses *dsolve* to solve the differential equation *eqn*. Note how the initial conditions are specified. D is a differential operator so that $D(r)(0) = 0$ is the initial condition $\dot{r} = 0$ at t =0.

> dsolve( {eqn, r(0)=r[0], D(r)(0)=0}, r(t));

$$\mathrm{r}(t) = \frac{1}{2} r_0 \, \mathrm{e}^{(\omega_0 t)} + \frac{1}{2} r_0 \, \mathrm{e}^{(-\omega_0 t)}$$

This expression is identical to that obtained in the problem formulation section above. Now we'll use polar coordinates to plot the path of the particle.

> r:=r0*cosh(theta);

$$r := r0 \cosh(\theta)$$

> r0:=0.1;

$$r0 := .1$$

> polarplot(r,theta=0..2.993, scaling=CONSTRAINED);

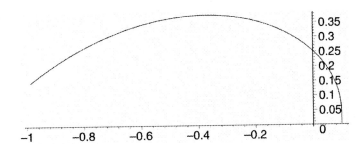

## 3.3 Sample Problem 3/17 (Potential Energy)

The 10-kg slider $A$ moves with negligible friction up the inclined guide. The attached spring has a stiffness of 60 N/m and is stretched $\delta$ m at position $A$, where the slider is released from rest. The 250-N force is constant and the pulley offers negligible resistance to the motion of the cord. Plot the velocity of the slider as it passes $C$ as a function of the initial spring stretch $\delta$. Let $\delta$ vary between $-0.4$ and $0.8$ m and explain the results when $\delta$ exceeds a value of about 0.65 m.

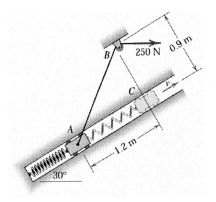

### Problem Formulation

The change in the elastic potential energy is

$$\Delta V_e = \frac{1}{2}k\left(x_2^2 - x_1^2\right) = \frac{1}{2}k\left((1.2 + \delta)^2 - \delta^2\right)$$

The other results in the sample problem (see your text) are unchanged,

$$U'_{1-2} = 250(0.6) = 150 \text{ J} \qquad \Delta T = \frac{1}{2}m\left(v^2 - v_0^2\right) = \frac{1}{2}(10)v^2$$

$$\Delta V_g = mg\Delta h = 10(9.81)(1.2 \sin 30) = 58.9 \text{ J}$$

$$U'_{1-2} = 150 = \frac{1}{2}(10)v^2 + 58.9 + \frac{1}{2}(60)\left((1.2 + \delta)^2 - \delta^2\right)$$

This equation can be solved for $v$ either by hand or by using Maple. The result is

$$v = \frac{1}{10}\sqrt{958 - 1440\delta}$$

### *Maple Worksheet*

> restart; with(plots):

> U:=150=1/2*10*v^2+58.9+1/2*60*((1.2+delta)^2-delta^2);

$$U := 150 = 5\,v^2 + 58.9 + 30\,(1.2 + \delta)^2 - 30\,\delta^2$$

> solve(U,v);

$$\frac{1}{10}\sqrt{958 - 1440\,\delta},\ -\frac{1}{10}\sqrt{958 - 1440\,\delta}$$

> v:=%[1];

$$v := \frac{1}{10}\sqrt{958 - 1440\,\delta}$$

> evalf(solve(v=0,delta));

.6652777778

> plot(v,delta=-0.4..0.8, labels=[`initial spring stretch (m)`,`velocity (m/s)`], labelfont=[TIMES,ROMAN,13]);

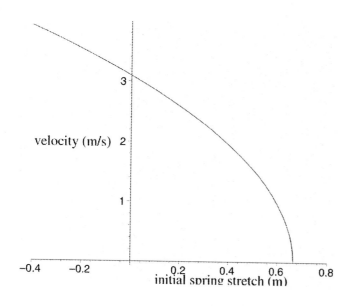

Note that no results are plotted beyond $\delta \cong 0.65$ m. Why? One reason is to observe from the above equation that $v$ becomes imaginary when $\delta > 958/1440 = 0.665$ m. No results are plotted because Maple does not plot imaginary numbers. But this is a numerical reason instead of a physical explanation. Usually, imaginary answers signify a situation that is physically impossible for some reason. One way of understanding this is as follows. If the spring is initially compressed it will, at least for some part of the motion, be pushing up and thus be aiding the 250 N force in overcoming the weight of the slider. If the spring is initially stretched, it will always be pulling back on the slider. Thus the 250 N force will have to overcome not only the weight but also the spring force. It stands to reason then that there will be some value for the initial spring stretch beyond which the 250 N force will not be able to pull the slider all the way to $C$. This value is found from the limiting case where $v = 0$. Thus, the block never reaches $C$ if $\delta > 0.665$ m.

---

## 3.4 Problem 3/218 (Linear Impulse/Momentum)

The ballistic pendulum is a simple device to measure the projectile velocity $v$ by observing the maximum angle $\theta$ to which the box of sand with embedded particle swings. As an aid for a laboratory technician, make a plot of the velocity $v$ in terms of the maximum angle $\theta$. Assume that the weight of the box is 50-lb while the weight of the projectile is 2-oz.

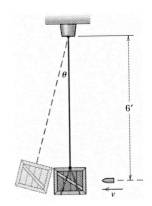

### Problem Formulation

*(1) Impulse/Momentum*

During impact, $\Delta G = 0$ and $G_1 = G_2$

$$\left(\frac{2/16}{32.2}\right)v + \left(\frac{50}{32.2}\right)(0) = \left(\frac{2/16 + 50}{32.2}\right)v_b$$

$$v = 401 v_b$$

where $v$ is the velocity of the projectile while $v_b$ is the velocity of the box of sand immediately after impact.

*(2)  Work/Energy*

Now we use the work/energy equation with our initial position being the position where the pendulum is still vertical ($\theta = 0$) and the final position is that where the pendulum has rotated through the maximum angle $\theta$.

$$U'_{1-2} = 0 = \Delta T + \Delta V_g = \frac{1}{2} m\left(0^2 - v_b^2\right) + mg\Delta h$$

where $m$ is the combined mass of the box and the projectile.

$$v_b = \sqrt{2g\Delta h} = \sqrt{2(32.2)(6)(1 - \cos\theta)}$$

$$v = 401 v_b = 7882\sqrt{1 - \cos\theta}$$

***Maple Worksheet***

> restart;
> vb:=sqrt(2*32.2*6*(1-cos(theta)));     # work/energy

$$vb := \sqrt{386.4 - 386.4\cos(\theta)}$$

> eqn:=2/16*v=(2/16+50)*vb;     # impulse/momentum

$$eqn := \frac{1}{8} v = \frac{401}{8}\sqrt{386.4 - 386.4\cos(\theta)}$$

> v:=solve(eqn,v);

$$v := 160.4000000\sqrt{2415. - 2415.\cos(\theta)}$$

Now we make a substitution yielding vd which is the velocity as a function of x where x is the angle theta in degrees.

> vd:=subs(theta=x*Pi/180,v);

$$vd := 160.4000000\sqrt{2415. - 2415.\cos\left(\frac{1}{180} x \pi\right)}$$

>

> plot(vd,x=0..90,labels=["theta (deg)",""],title="velocity of bullet (ft/s)");

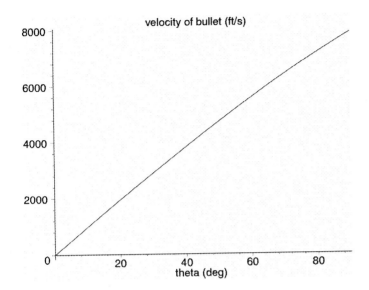

velocity of bullet (ft/s)

## 3.5 Problem 3/250 (Angular Impulse/Momentum)

The assembly of two 5-kg spheres is rotating freely about the vertical axis at 40 rev/min with $\theta = 90°$. The force $F$ that maintains the given position is increased to raise the base collar and reduce the angle from 90° to an arbitrary angle $\theta$. Determine the new angular velocity $\omega$ and plot $\omega$ as a function of $\theta$ for $0 \le \theta \le 90°$. Assume that the mass of the arms and collars is negligible.

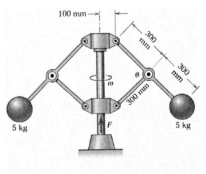

### Problem Formulation

Since the summation of moments about the vertical axis is zero we have conservation of angular momentum about that axis. The spheres are rotating in a circular path about the vertical axis. The angular momentum of a particle moving in a circular path of radius $r$ with angular velocity $\omega$ is $H = mr^2\omega$. Thus, from the conservation of angular momentum we have,

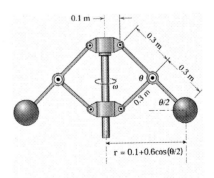

$$2mr_0^2 \omega_0 = 2mr^2 \omega \qquad \omega = \frac{r_0^2}{r^2} \omega_0$$

where $\qquad r_0 = 0.1 + 0.6\cos(45^0) \quad$ and $\qquad r = 0.1 + 0.6\cos(\theta/2)$

**Maple Worksheet**

> restart;
> r0:=0.1+0.6*cos(Pi/4);
$$r0 := .1 + .3000000000\sqrt{2}$$

> r:=0.1+0.6*cos(theta/2);
$$r := .1 + .6\cos\left(\frac{1}{2}\theta\right)$$

> omega:=r0^2/r^2*omega0;
$$\omega := \frac{\left(.1 + .3000000000\sqrt{2}\,\right)^2 \omega0}{\left(.1 + .6\cos\left(\frac{1}{2}\theta\right)\right)^2}$$

> omega:=subs(theta=x*Pi/180,omega);
$$\omega := \frac{\left(.1 + .3000000000\sqrt{2}\,\right)^2 \omega0}{\left(.1 + .6\cos\left(\frac{1}{360}x\,\pi\right)\right)^2}$$

The above converts $\omega$ to a function of x where x is the angle $\theta$ in degrees instead of radians.

> omega0:=40*2*Pi/60;
$$\omega0 := \frac{4}{3}\pi$$

> plot(omega,x=0..90,labels=["theta (degrees)","omega (rad/s)"],labeldirections=[HORIZONTAL,VERTICAL]);

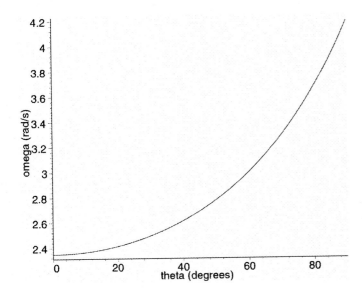

This diagram "begins" at $\theta = 90°$ where $\omega = \omega_0 = 40(2\pi)/60 = 4.19$ rad/s.

### 3.6 Problem 3/365 (Curvilinear Motion)

The 26-in. drum rotates about a horizontal axis with a constant angular velocity $\Omega = 7.5$ rad/sec. The small block $A$ has no motion relative to the drum surface as it passes the bottom position $\theta = 0$. Determine the coefficient of static friction $\mu_s$ that would result in block slippage at an angular position $\theta$, plot your expression for $0 \leq \theta \leq 180°$. Determine the minimum required coefficient value $\mu_{min}$ that would allow the block to remain fixed relative to the drum throughout a full revolution. For a friction coefficient slightly less than $\mu_{min}$, at what angular position $\theta$ would slippage occur?

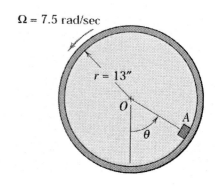

### Problem Formulation

From the free body and mass acceleration diagrams we have,

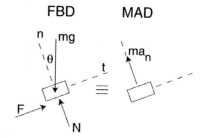

$$[\Sigma F_n = ma_n] \qquad N - mg\cos\theta = mr\Omega^2$$

$$[\Sigma F_t = ma_t] \qquad F - mg\sin\theta = 0$$

For impending slip we have $F = \mu_s N$. Substituting F into the above and solving gives,

$$\mu_s = \frac{g\sin\theta}{g\cos\theta + r\Omega^2} = \frac{\sin\theta}{1.8925 + \cos\theta}$$

The last two questions can be answered only after plotting $\mu_s$ as a function of $\theta$.

### Maple Worksheet

```
> restart;
> eqn1:=N-m*g*cos(theta)=m*r*Omega^2;
        eqn1 := N - m g cos(θ) = m r Ω²

> eqn2:=mu[s]*N-m*g*sin(theta)=0;
        eqn2 := μ_s N - m g sin(θ) = 0

> soln:=solve({eqn1,eqn2},{mu[s],N});
```

$$soln := \{ N = m\,g\,\cos(\theta) + m\,r\,\Omega^2, \; \mu_s = \frac{g\,\sin(\theta)}{g\,\cos(\theta) + r\,\Omega^2} \}$$

> assign(soln):

> Omega:=7.5: g:=32.2: r:=13/12:

> plot(mu[s],theta=0..Pi,color=black,labels=["theta (rads)",""],title="coefficient of static friction");

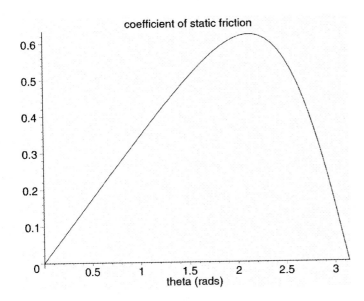

If the block is not to slip at any angle $\theta$, the coefficient of friction must be greater than or equal to any value shown on the plot above. Thus, the minimum required coefficient value $\mu_{min}$ that would allow the block to remain fixed relative to the drum throughout a full revolution is equal to the maximum value in the plot above. The location where this maximum occurs can be found by solving the equation $d\mu_s / d\theta = 0$ for $\theta$. This $\theta$ can then be substituted into $\mu_s$ to yield the required value for $\mu_{min}$. Here's how we do this with Maple.

> dmu:=diff(mu[s],theta);

$$dmu := 32.2\frac{\cos(\theta)}{32.2\cos(\theta) + 60.93750000} + \frac{1036.84\sin(\theta)^2}{(32.2\cos(\theta) + 60.93750000)^2}$$

> solve(dmu=0,theta);

$$-2.127523285, \, 2.127523285$$

> evalf(subs(theta=2.12752,mu[s]));
                .6223992940

From the above we see that $\mu_{min} = 0.622$. If $\mu_s$ is slightly less than this value, the block will slip when $\theta = 2.128$ rads (121.9°).

# KINETICS OF SYSTEMS OF PARTICLES

# 4

This chapter concerns the extension of principles covered in chapters two and three to the study of the motion of general systems of particles. The chapter first considers the three approaches introduced in chapter 3 (direct application of Newton's second law, work/energy, and impulse/momentum) and then moves to other applications such as steady mass flow and variable mass. Problem 4.1 considers an application of the conservation of momentum to a system comprised of a small car and an attached rotating sphere. Maple is used to plot the velocity of the car as a function of the angular position of the sphere. The absolute position of the sphere is also plotted. Problem 4.2 uses the concept of steady mass flow to study the effects of geometry upon the design of a sprinkler system. One of the main purposes of this problem is to illustrate how a problem can be greatly simplified using non-dimensional analysis. In particular, an equation containing seven parameters is reduced to a non-dimensional equation with only three parameters. Problem 4.3 is a variable mass problem in which Maple is used to integrate the kinematic equation $vdv = adx$.

## 4.1 Problem 4/26 (Conservation of Momentum)

The small car, which has a mass of 20 kg, rolls freely on the horizontal track and carries the 5-kg sphere mounted on the light rotating rod with $r = 0.4$ m. A geared motor drive maintains a constant angular speed $\dot{\theta} = 4$ rad/s of the rod. If the car has a velocity $v = 0.6$ m/s when $\theta = 0$, plot $v$ as a function of $\theta$ for one revolution of the rod. Also plot the absolute position of the sphere for two revolutions of the rod. Neglect the mass of the wheels and any friction.

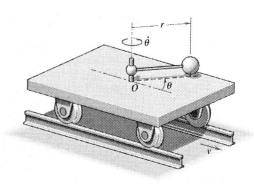

### Problem Formulation

Since $\Sigma F_x = 0$ we have conservation of momentum in the $x$ direction. The diagram to the right shows the system at $\theta = 0$ and at an arbitrary angle $\theta$. From the relative velocity equation, the velocity of the sphere is the vector sum of the velocity of the car ($v$) and the velocity of the sphere relative to the car ($r\dot{\theta}$).

$$(G_x)_{\theta=0} = 20(0.6) + 5(0.6) = 15 \text{ N}\cdot\text{s}$$

$$(G_x)_{\theta} = 20v + 5(v - r\dot{\theta}\sin\theta) = 25v - 8\sin\theta$$

Setting $(G_x)_{\theta=0} = (G_x)_{\theta}$ and solving yields,

$$v = 0.6 + 0.32\sin\theta$$

Now let time $t = 0$ be the time when $\theta = 0$ and place an $x$-$y$ coordinate system at the center of the car as shown in the diagram so that $x(t)$ is the position of the center of the car. Since $v = dx/dt$ and $\theta = 4t$ we have,

$$x = \int_0^t v\,dt = \int_0^t (0.6 + 0.32\sin(4t))dt = 0.6t + 0.08(1 - \cos(4t))$$

The $x$ and $y$ components of the sphere can now be determined as,

$$x_s = x + r\cos\theta = 0.08 + 0.6t + 0.32\cos(4t)$$

$$y_s = r \sin \theta = 0.4 \sin(4t)$$

The absolute position of the sphere can be obtained by plotting $y_s$ versus $x_s$. The time required for two revolutions of the arm is $4\pi/4 = \pi$ seconds.

## Maple Worksheet

```
> restart;
> v:=0.6+0.32*sin(theta);
```
$$v := .6 + .32 \sin(\theta)$$

```
> plot([theta*180/Pi,v,theta=0..2*Pi],labels=["theta (degrees)","v (m/s)"], labeldirections
=[HORIZONTAL,VERTICAL]);
```

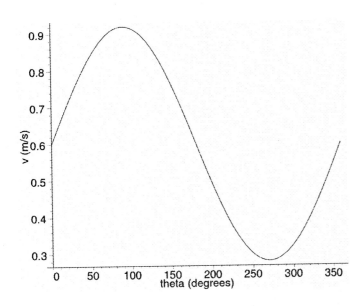

```
> theta:=4*t;
```
$$\theta := 4t$$

```
> x:=int(subs(t=q,v),q=0..t);
```
$$x := .6000000000t - .08000000000\cos(4.t) + .08000000000$$

```
> xs:=x+0.4*cos(theta);
```
$$xs := .6000000000t - .08000000000\cos(4.t) + .08000000000 + .4\cos(4t)$$

```
> ys:=0.4*sin(theta);
```
$$ys := .4\sin(4t)$$

```
> tf:=solve(theta=4*Pi,t);
```

$$tf := \pi$$

> plot([xs,ys,t=0..tf],scaling=CONSTRAINED,labels=["x (m)","y(m)"],
labeldirections =[HORIZONTAL,VERTICAL]);

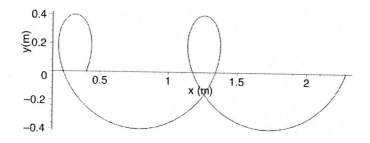

## 4.2 Problem 4/62 (Steady Mass Flow)

The sprinkler is made to rotate at the constant angular velocity $\omega$ and distributes water at the volume rate $Q$. Each of the four nozzles has an exit area $A$. Write an expression for the torque $M$ on the shaft of the sprinkler necessary to maintain the given motion. Here we would like to study the effects of the geometry of the sprinkler upon this torque. To this end, it is helpful to introduce the non-dimensional parameters $M' = M/4\rho A r u^2$, $\Omega = \omega r/u$, and $\beta = b/r$ where $u = Q/4A$ is the velocity of the water relative to the nozzle and $\rho$ is the density of the water. Plot the non-dimensional torque $M'$ versus $\beta$ for $\Omega = 0.5$, 1, and 2. Let $\Omega_0$ be the non-dimensional velocity $\Omega$ at which the sprinkler will operate with no applied torque. Plot $\Omega_0$ versus $\beta$. For both plots let $\beta$ range between 0 and 1.

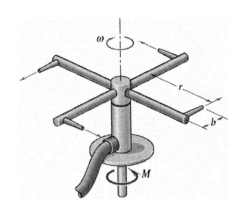

### Problem Formulation

The figure to the right shows the three components of the absolute velocity of the water at the exit. $u \ (= Q/4A)$ is the velocity of the water relative to the nozzle. The mass flow rate $m' = \rho Q$. Taking clockwise as positive, the application of equation 4/19 of your text yields,

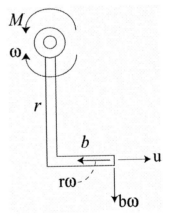

$$\Sigma M_0 = -M = \rho Q\left(\omega r^2 + \omega b^2 - ur - 0\right)$$

$$M = \rho Q\left(ur - \omega\left(r^2 + b^2\right)\right)$$

Now we want to introduce the non-dimensional parameters defined in the problem statement. For many undergraduate students, non-dimensional analysis is a very confusing topic. It is important to realize that the difficulty is really that of determining which non-dimensional parameters are appropriate for a particular problem. If these parameters have already been defined, as in this problem, all you have to do is substitute. In this case we merely substitute $M = 4\rho A r u^2 M'$, $\omega = \Omega u/r$, and $b = r\beta$ into the equation above. When this is done many terms will cancel yielding,

$$M' = 1 - \Omega\left(1 + \beta^2\right)$$

Setting $M' = 0$ we can solve for $\Omega_0$,

$$\Omega_0 = \frac{1}{1 + \beta^2}$$

### *Maple Worksheet*

```
> restart; with(plots):
> Mp:=1-Omega*(1+beta^2);
          Mp := 1 − Ω ( 1 + β² )

> Mp1:=subs(Omega=0.5,Mp): Mp2:=subs(Omega=1,Mp):
> Mp3:=subs(Omega=2,Mp):
> p1:=plot([Mp1,Mp2,Mp3],beta=0..1,labels=["b/r",""], title=
"non-dimensional torque"):
> t1:=textplot([0.72,0.4,"W = 0.5"],font=[SYMBOL,12]):
> t2:=textplot([0.7,-0.3,"W = 1"],font=[SYMBOL,12]):
> t3:=textplot([0.7,-1.7,"W = 2"],font=[SYMBOL,12]):
> display(p1,t1,t2,t3);
```

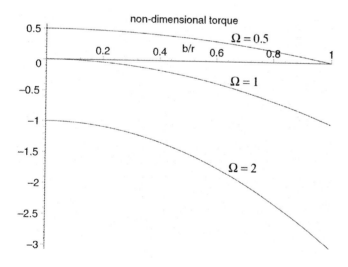

```
> Omega[0]:=1/(1+beta^2);
              Ω₀ := \frac{1}{1 + β²}

> plot(Omega[0],beta=0..1,labels=["b","W0"],labelfont=[SYMBOL,14]);
```

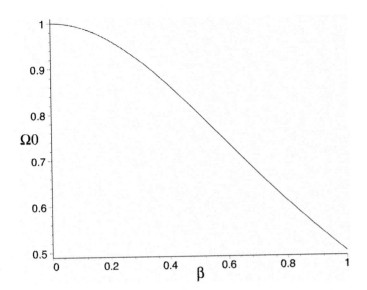

---

## 4.3 Problem 4/86 (Variable Mass)

The open-link chain of length $L$ and mass $\rho$ per unit length is released from rest in the position shown, where the bottom link is almost touching the platform and the horizontal section is supported on a smooth surface. Friction at the corner guide is negligible. Determine (a) the velocity $v_1$ of end $A$ as it reaches the corner and (b) its velocity $v_2$ as it strikes the platform. Plot $v_1$ and $v_2$ as functions of $h$ for $L = 5$ m.

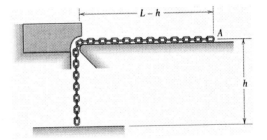

### Problem Formulation

Let $x$ be the displacement of the chain and $T$ be the tension in the chain at the corner as shown in the diagram to the right. Note that the acceleration of the horizontal and vertical sections are both equal to $\ddot{x}$.

For the horizontal section,

$$\left[\Sigma F_x = ma_x\right] \qquad T = \rho(L - h - x)\ddot{x}$$

For the vertical section,

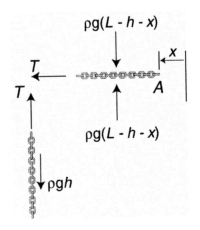

$$\downarrow \left[ \Sigma F_y = ma_y \right] \qquad \rho g h - T = \rho h \ddot{x}$$

Substituting $T$ from the first equation into the second and simplifying gives,

$$\ddot{x} = \frac{gh}{L - x}$$

Now we use the relation $vdv = adx$ to write,

$$\int_0^{v_1} vdv = \int_0^{L-h} \frac{gh}{L - x} dx \qquad v_1^2 = 2 \int_0^{L-h} \frac{gh}{L - x} dx = 2gh \ln\left(\frac{L}{h}\right)$$

$$v_1 = \sqrt{2gh \ln(L/h)}$$

After end $A$ has passed the corner it will be in free-fall until it hits the platform. With $y$ positive down we have $vdv = gdy$ yielding,

$$\frac{1}{2}\left(v_2^2 - v_1^2\right) = gh$$

Substituting for $v_1$ and solving,

$$v_2 = \sqrt{2gh(1 + \ln(L/h))}$$

### *Maple Worksheet*

```
> restart; with(plots):
> v1:=sqrt(2*g*h*log(L/h));
```
$$v1 := \sqrt{2}\sqrt{g\,h\,\ln\left(\frac{L}{h}\right)}$$

```
> v2:=sqrt(2*g*h*(1+log(L/h)));
```
$$v2 := \sqrt{2}\sqrt{g\,h\left(1 + \ln\left(\frac{L}{h}\right)\right)}$$

```
> L:=5:g:=9.81:
> p:=plot([v1,v2],h=0..L, labels=["h (m)",""],title="velocity (m/s)"):
> t1:=textplot([2,6.5,"v1"]): t2:=textplot([2,9.2,"v2"]):

> display(p,t1,t2);
```

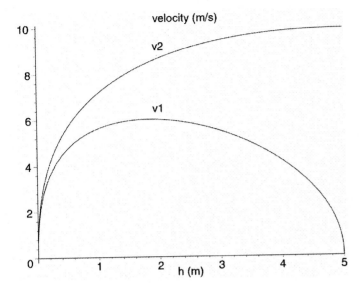

# PLANE KINEMATICS OF RIGID BODIES

# 5

This chapter extends the kinematic analysis of particles covered in chapter 2 to rigid bodies by taking into account the rotational motion of the body. Thus, the motion of rigid bodies involves both translation and rotation. Problem 5.1 contrasts net angular displacement and total number of cycles. The problem also introduces the *piecewise* function. Problem 5.2 is an interesting application of absolute motion analysis. Illustrated in this problem are the simultaneous (symbolic) solution of multiple equations and differentiation with respect to time of a function whose explicit time dependence is not known. In problem 5.3 the velocity of the piston in a reciprocating engine is plotted versus the angular orientation of the crank. The maximum velocity and the corresponding orientation (of the crank) are obtained using *diff* and *fsolve*. This is a rare case where *fsolve* is preferred over *solve* for reasons discussed in the problem. The instantaneous center of zero velocity is used in problem 5.4 to relate the velocity of a vertical control rod to the angular velocity of a rotating bar in a switching device. Problem 5.5 is a fairly straightforward relative acceleration problem while problem 5.6 considers the reciprocating engine of problem 5.3 but uses absolute rather than relative motion analysis.

## 5.1 Problem 5/3 (Rotation)

The angular velocity of a gear is controlled according $\omega = 12 - 3t^2$ where $\omega$, in radians per second, is positive in the clockwise sense and where $t$ is the time in seconds. Find the net angular displacement $\Delta\theta$ and the total number of revolutions $N$ through which the gear turns in terms of the time $t$. Plot $\Delta\theta$ and $N$ for time $t$ up to 4 seconds.

### Problem Formulation

You may want to have a look at sample problem 5/1 in your text before continuing with this problem. In particular it is important to note the difference between the angular displacement $\Delta\theta$ and total number of revolutions $N$. The angular displacement is the integral of $\omega$ over time and can be positive or negative depending on whether the rotation is clockwise or counterclockwise. Referring to the graph of $\omega$ to the right, $\Delta\theta$ will be the total area under the curve up to a particular time. The area is negative when the curve dips below

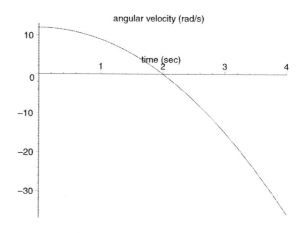

the time axis. As the name implies, the total number of revolutions simply counts the number of times the disk rotates and is not concerned with whether the rotation is clockwise or counterclockwise. Thus, $N$ will be proportional to the magnitude of the area under the angular velocity curve. For example, suppose the gear rotates two revolutions clockwise followed by two revolutions counterclockwise. In this case $\Delta\theta = 0$ but $N = 4$.

Based upon the above it is essential to know beforehand the time when $\omega = 0$ (i.e. when the gear changes diection) in order to calculate $N$. Setting $12 - 3t^2 = 0$ gives $t = 2$ seconds. For time greater than 2 sec. we need to break the integral up into two parts when calculating $N$. For $\Delta\theta$ we do not need to keep track of when the gear changes directions. Thus,

$$\left[ \omega = \frac{d\theta}{dt} \right] \qquad \Delta\theta = \int_0^t \omega dt = \int_0^t \left(12 - 3u^2\right) du = 12t - t^3$$

for t < 2 sec

$$N = \frac{1}{2\pi} \int_0^t \omega \, dt = \int_0^t (12 - 3t^2) \, dt = \frac{1}{2\pi} (12t - t^3)$$

for t > 2 sec

$$N = \frac{1}{2\pi} \int_0^2 \omega \, dt - \frac{1}{2\pi} \int_2^t \omega \, dt = \frac{1}{2\pi} \left( \int_0^2 (12 - 3t^2) \, dt - \int_2^t (12 - 3t^2) \, dt \right)$$

$$N = \frac{1}{2\pi} (32 - 12t + t^3)$$

In summary,

$$\Delta\theta = 12t - t^3 \qquad \text{all } t$$

$$N = \begin{cases} \dfrac{1}{2\pi}(12t - t^3) & t < 2\sec \\ \dfrac{1}{2\pi}(32 - 12t + t^3) & t > 2\sec \end{cases}$$

Although the expression for $N$ is rather simple, plotting the result can be a little tricky due to the change in the expression at time $t = 2$ sec. How do we tell Maple to stop plotting one expression and start plotting the other?

Here we will use two approaches to handle this difficulty. The first is to use Maple's *piecewise* function. This function can be used whenever a function has different expressions depending upon the value of the independent variable. The format is illustrated below for our function $N$.

> N:=1/2/Pi*piecewise(t<2,12*t-t^3,t>2,32-12*t+t^3);

$$N := \frac{1}{2} \frac{ \begin{cases} 12\,t - t^3 & t < 2 \\ 32 - 12\,t + t^3 & 2 < t \end{cases} }{\pi}$$

Note that we simply write the constraint followed by the expression. Any number of constraints can be listed.

The second approach allows to do almost all of the work within Maple. This is accomplished first by observing that we can obtain the *magnitude* of the area beneath the $\omega$ curve by integrating the absolute value of $\omega$.

$$N = \frac{1}{2\pi} \int_0^t |\omega| dt \qquad \text{for all } t$$

As we shall see below, Maple automatically turns this integral into a piecewise function.

### *Maple Worksheet*

```
> restart;
> omega:=12-3*u^2;
```
$$\omega := 12 - 3\ u^2$$

```
> plot(omega,u=0..4,labels=["time (sec)",""],title="angular velocity (rad/s)");
```

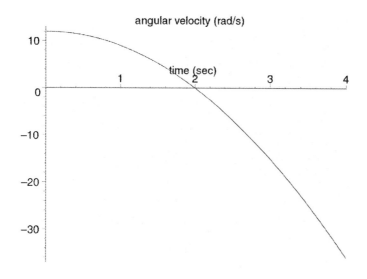

```
> theta:=int(omega,u=0..t);
```
$$\theta := 12\ t - t^3$$

Here we use the second approach mentioned in the problem formulation above. Remember that this approach does not require dividing the domain into two integrals for $t > 2$.

```
> N:=1/2/Pi*int(abs(omega),u=0..t);
```

$$N := \frac{1}{2}\ \frac{-32 + \left(\begin{cases} -12\ t + t^3 & t \le -2 \\ 12\ t - t^3 + 32 & t \le 2 \\ -12\ t + t^3 + 64 & 2 < t \end{cases}\right)}{\pi}$$

Note that Maple has constructed a piecewise function with an extra region, t < -2 sec. This corresponds to the second root of the equation $\omega = 12 - 3t^2 = 0$. We neglected this solution in the problem formulation section since it is not physically meaningful.

> plot(theta,t=0..4,labels=["time (sec)",""],title="angular position theta (degrees)");

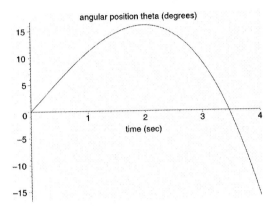

> plot(N,t=0..4,labels=["time (sec)",""],title="total number of revolutions N");

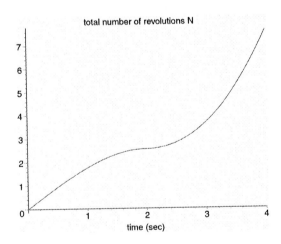

The following illustrates the use of the piecewise function.

> N:=1/2/Pi*piecewise(t<2,12*t-t^3,t>2,32-12*t+t^3);

$$N := \frac{1}{2} \frac{\begin{cases} 12\,t - t^3 & t < 2 \\ 32 - 12\,t + t^3 & 2 < t \end{cases}}{\pi}$$

> plot(N,t=0..4,labels=["time (sec)",""],title="total number of revolutions N");

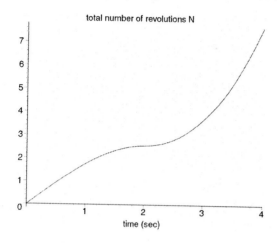

total number of revolutions N

## 5.2 Problem 5/44 (Absolute Motion)

Derive an expression for the upward velocity $v$ of the car hoist system in terms of $\theta$. The piston rod of the hydraulic cylinder is extending at the rate $\dot{s}$. Plot the non-dimensional velocity $v/\dot{s}$ as a function of $\theta$ for $b/L = 0.1, 0.5, 1$, and $2$.

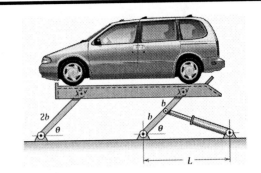

### Problem Formulation

From the diagram to the right,

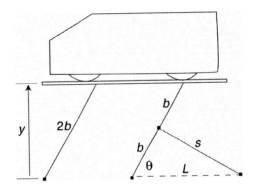

$$y = 2b\sin\theta$$

$$\dot{y} = 2b\dot{\theta}\cos\theta$$

If the angular velocity $\dot{\theta}$ were known as a function of $\theta$ we would be finished. The motion of the car hoist system is controlled by the extension rate $\dot{s}$ of the hydraulic cylinder rather than the angular velocity. Thus, to complete the problem we need to relate $\dot{\theta}$ and $\dot{s}$.

$$s^2 = L^2 + b^2 - 2Lb\cos\theta$$

$$2s\dot{s} = 0 + 0 + 2Lb\dot{\theta}\sin\theta$$

$$\dot{\theta} = \frac{s\dot{s}}{Lb\sin\theta}$$

Substituting,

$$v = \frac{2bs\dot{s}}{Lb\sin\theta}\cos\theta = \frac{2\dot{s}\sqrt{L^2 + b^2 - 2Lb\cos\theta}}{L\tan\theta}$$

$$v/\dot{s} = \frac{2\sqrt{1 + (b/L)^2 - 2(b/L)\cos\theta}}{\tan\theta} = \frac{2\sqrt{1 + \beta^2 - 2\beta\cos\theta}}{\tan\theta}$$

where $\beta = b/L$.

### Maple Worksheet

The algebra in this problem is relatively simple and hardly requires Maple's symbolic abilities. We will go ahead and solve the problem symbolically anyway for purposes of illustration.

```
> restart; with(plots):
> eqn1:=y(t)=2*b*sin(theta(t));
```
$$eqn1 := y(t) = 2\, b\, \sin(\theta(t))$$

```
> eqn2:=s(t)^2=L^2+b^2-2*L*b*cos(theta(t));
```
$$eqn2 := s(t)^2 = L^2 + b^2 - 2\, L\, b\, \cos(\theta(t))$$

Since we will differentiate with respect to t we need to tell Maple which variables depend on t. Thus, we write $\theta(t)$, $y(t)$, and $s(t)$ in the equations above.

```
> eqn1:=diff(eqn1,t); eqn2:=diff(eqn2,t);
```
$$eqn1 := \frac{d}{dt}\, y(t) = 2\, b\, \cos(\theta(t)) \left( \frac{d}{dt}\, \theta(t) \right)$$

$$eqn2 := 2\, s(t) \left( \frac{d}{dt}\, s(t) \right) = 2\, L\, b\, \sin(\theta(t)) \left( \frac{d}{dt}\, \theta(t) \right)$$

For convenience we will now make a few substitutions in order to remove the explicit dependence upon t.

```
> eqn1:=subs(diff(y(t),t)=v,diff(theta(t),t)=omega, theta(t)=theta,eqn1);
```
$$eqn1 := v = 2\, b\, \cos(\theta)\, \omega$$

```
> eqn2:=subs(diff(s(t),t)=sdot, s(t)=s,diff(theta(t),t)=omega, theta(t)=theta,eqn2);
```
$$eqn2 := 2\, s\, sdot = 2\, L\, b\, \sin(\theta)\, \omega$$

```
> s:=sqrt(L^2+b^2-2*L*b*cos(theta));
```
$$s := \sqrt{L^2 + b^2 - 2\, L\, b\, \cos(\theta)}$$

```
> solve({eqn1,eqn2},{v,omega});
```
$$\left\{ \omega = \frac{\sqrt{L^2 + b^2 - 2\, L\, b\, \cos(\theta)}\; sdot}{L\, b\, \sin(\theta)},\; v = 2\, \frac{\cos(\theta)\, \sqrt{L^2 + b^2 - 2\, L\, b\, \cos(\theta)}\; sdot}{L\, \sin(\theta)} \right\}$$

> v_nd:=2/tan(theta)*sqrt(1^2+beta^2-2*beta*cos(theta));

$$v\_nd := 2\,\frac{\sqrt{1 + \beta^2 - 2\,\beta\,\cos(\theta)}}{\tan(\theta)}$$

> v1:=subs(beta=0.1,v_nd):  v2:=subs(beta=0.5,v_nd):
> v3:=subs(beta=1,v_nd):    v4:=subs(beta=2,v_nd):
> p:=plot([v1,v2,v3,v4],theta=10*Pi/180..Pi/2,y=0..5,labels=["theta (rads)",""],
title="non-dimensional velocity"):
> t1:=textplot([0.8,3.4,"b=2"],font=[SYMBOL,10]):
> t2:=textplot([0.3,2.2,"b=1"],font=[SYMBOL,10]):
> t3:=textplot([0.62,3,"0.1"],font=[SYMBOL,10]):
> t4:=textplot([0.48,2.65,"0.5"],font=[SYMBOL,10]):
>
> display(p,t1,t2,t3,t4);

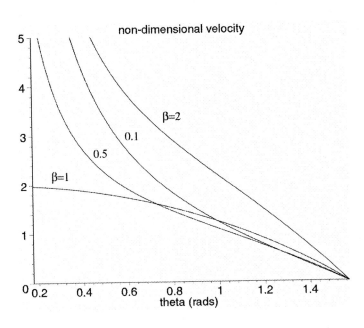

## 5.3 Sample Problem 5/9 (Relative Velocity)

The common configuration of a reciprocating engine is that of the slider crank mechanism shown. If crank $OB$ has a clockwise rotational speed of 1500 rev/min; (a) Plot $v_A$ versus $\theta$ for one revolution of the crank. (b) Find the maximum speed of the piston $A$ and the corresponding value of $\theta$.

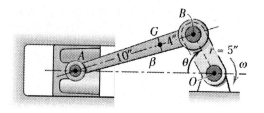

### Problem Formulation

Let $l$ be the length of connecting rod $AB$ and start with the relative velocity equation,

$$\mathbf{v}_B = \mathbf{v}_A + \mathbf{v}_{BA}$$

The crank pin velocity is $v_B = r\omega$ and is normal to $OB$. The velocity of $A$ is horizontal while the velocity of $B/A$ has magnitude $l\omega_{AB}$ and is directed perpendicular to $AB$. The angle $\beta$ can be found in terms of $\theta$ by using the law of sines,

$$\sin \beta = \frac{r}{l} \sin \theta \quad \text{Also,} \quad \cos \beta = \sqrt{1 - \sin^2 \beta} = \sqrt{1 - \frac{r^2}{l^2} \sin^2 \theta}$$

From the vector diagram to the right,

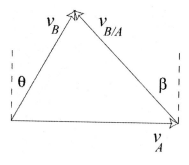

$$v_B \cos \theta = v_{B/A} \cos \beta . \text{ Thus, } v_{B/A} = \frac{\cos \theta}{\cos \beta} r\omega$$

Also from the diagram, $v_A = v_B \sin \theta + v_{B/A} \sin \beta$.
Substitution yields

$$v_A = r\omega(\sin \theta + \cos \theta \tan \beta) = r\omega \sin \theta \left( 1 + \frac{r \cos \theta}{l\sqrt{1 - \frac{r^2}{l^2} \sin^2 \theta}} \right)$$

Note that $v_A$ has been expressed explicitly in terms of $\theta$ by substituting for $\cos\beta$ and $\tan\beta$. This has been done only for sake of clarity. When working with a computer such substitutions will be automatic once $\beta$ has been defined in terms of $\theta$ ( $\beta = \sin^{-1}(r\sin\theta/l)$ ).

(b) The angle $\theta$ at which the maximum value of $v_A$ occurs is found by solving the equation $dv_A/d\theta = 0$ for $\theta$. Evaluating this derivative and solving the resulting equation for $\theta$ would be difficult without the help of a computer. It turns out that there are many solutions to this equation, most of which are complex. In the results that follow we find that the maximum occurs at $\theta = 1.261$ radians (72.3°). Substitution of this result into the expression for $v_A$ yields the maximum speed of the piston $A$, $v_A = 69.6$ ft/sec.

### Maple Worksheet

> restart; with(plots):

> vB:=r*omega;
$$vB := r\,\omega$$

> beta:=arcsin(r/L*sin(theta));
$$\beta := \arcsin\left(\frac{r\sin(\theta)}{L}\right)$$

> vBA:=cos(theta)/cos(beta)*r*omega;
$$vBA := \frac{\cos(\theta)\,r\,\omega}{\sqrt{1 - \frac{r^2\sin(\theta)^2}{L^2}}}$$

> vA:=vB*sin(theta)+vBA*sin(beta);
$$vA := r\,\omega\sin(\theta) + \frac{\cos(\theta)\,r^2\,\omega\sin(\theta)}{\sqrt{1 - \frac{r^2\sin(\theta)^2}{L^2}}\,L}$$

> L:=14/12: r:=5/12: omega:=1500*2*Pi/60:
>plot(vA,theta=0..2*Pi,labels=[`theta(radians)`, vA(ft/sec)`], labeldirections=[HORIZONTAL,VERTICAL]);

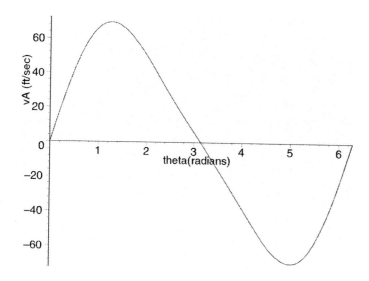

Part (b)

> dvAdtheta:=diff(vA,theta);

*dvAdtheta* :=

$$\frac{125}{6}\,\pi\cos(\theta) - \frac{625}{84}\,\frac{\sin(\theta)^2\,\pi}{\sqrt{1 - \frac{25}{196}\sin(\theta)^2}} + \frac{\frac{15625}{16464}\cos(\theta)^2\,\pi\sin(\theta)^2}{\left(1 - \frac{25}{196}\sin(\theta)^2\right)^{(3/2)}} + \frac{\frac{625}{84}\cos(\theta)^2\,\pi}{\sqrt{1 - \frac{25}{196}\sin(\theta)^2}}$$

> fsolve(dvAdtheta = 0, theta);
          1.261203789

Here we see only one of the many roots mentioned above. The reason is that we
have used *fsolve* (instead of *solve*), which has in this case returned only the first
real root. Try using *solve* and you will find about six solutions. Each one is a
lengthy symbolic result so the list of solutions will take several pages. This is the
main reason it was omitted here.

Although *fsolve* is very convenient in that it avoids lengthy symbolic results, you
also have to be careful when there are multiple solutions since it may not return
the solution of interest. For this reason, you should always produce a plot of the
function in order to insure that you have found the appropriate solution.
Referring to the plot above we see that we have found the value of theta
corresponding to the maximum speed. If you fail to find the correct solution you

should repeat *fsolve* specifying a range. For example, suppose you wanted the angle at which the speed is a minimum. From the plot we see that this occurs at about 5 radians.

> fsolve(dvAdtheta = 0, theta,theta=4..6);
           5.021981518

Now we can find the maximum velocity by substitution.

> evalf(subs(theta=1.2612,vA));
           69.55144676

Thus, the maximum speed of the piston is $v_A = 69.6$ ft/sec at $\theta = 1.261$ rad (72.3°).

---

## 5.4 Problem 5/108 (Instantaneous Center)

The switching device of Prob. 5/85 is repeated here. If the vertical control rod has a constant downward velocity $v$ of 3 ft/sec and if roller $A$ is in continuous contact with the horizontal surface, determine by the method of this article the magnitude of the velocity of $A$ and of $C$ as functions of $\theta$. Plot $v_A$ and $v_C$ for $\theta$ between 20 and 70°.

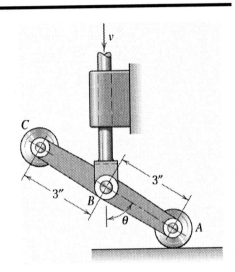

### Problem Formulation

The instantaneous center of zero velocity for the link is shown to the right. From right triangle $COC_1$ we can find the length

$$CC_1 = \sqrt{(6\sin\theta)^2 + (3\cos\theta)^2}$$

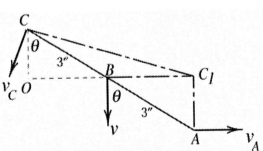

Using the instantaneous center we can now write,

$$v_B = 3\sin\theta \qquad \omega = \frac{v_B}{3\sin\theta} = \frac{v}{3\sin\theta}$$

$$v_A = 3\cos\theta\,\omega = 3\cos\theta\frac{v}{3\sin\theta} \qquad v_A = \frac{v}{\tan\theta}$$

$$v_C = CC_1 \omega = \frac{CC_1}{3 \sin \theta} v$$

We will let the computer substitute for $CC_1$ as usual. Also note that $v$ must be in in./sec.

### Maple Worksheet

> restart; with(plots):
> v:=3*12;

$$v := 36$$

> CC1:=sqrt((6*sin(theta))^2+(3*cos(theta))^2);

$$CC1 := 3 \sqrt{4 \sin(\theta)^2 + \cos(\theta)^2}$$

> vA:=v/tan(theta);

$$vA := \frac{36}{\tan(\theta)}$$

> vC:=CC1/3/sin(theta)*v;

$$vC := \frac{36 \sqrt{4 \sin(\theta)^2 + \cos(\theta)^2}}{\sin(\theta)}$$

> p1:=plot([vA/12,vC/12], theta=20*Pi/180..70*Pi/180, color=black, labels=["theta (rads)",""], title="velocity (ft/sec)"):
> t1:=textplot([.8,3.3,"vA"]):
> t2:=textplot([.8,7,"vC"]):
> display(p1,t1,t2);

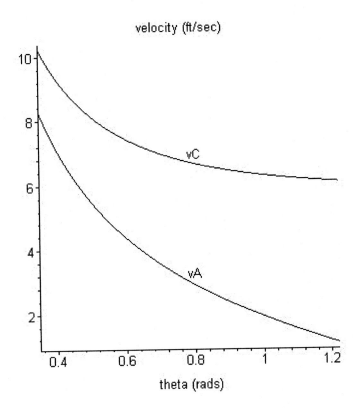

## 5.5 Problem 5/123 (Relative Acceleration)

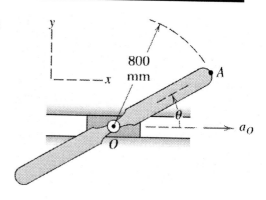

The two rotor blades of radius $r$ = 800-mm rotate counterclockwise with a constant angular velocity $\omega$ about the shaft at $O$ mounted in the sliding block. The acceleration of the block is $a_O$. Determine the magnitude of the acceleration of the tip $A$ of the blade in terms of $r$, $\omega$, $a_O$, and $\theta$. Plot the acceleration of $A$ as a function of $\theta$ for one revolution if $a_O$ = 3 m/s. Consider three cases: $\omega$ = 2, 4, and 6 rad/s.

**Problem Formulation**

The acceleration of $A$ relative to $O$ is

$$\vec{a}_A = \vec{a}_O + (\vec{a}_{A/O})_n + (\vec{a}_{A/O})_t$$

The acceleration of $O$ is to the right while the normal relative acceleration must point from $A$ towards $O$. Since $\omega$ is constant, the tangential relative acceleration will be zero. These considerations lead to the vector diagram shown to the right. Using the law of cosines,

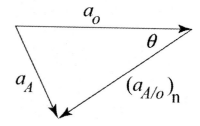

$$a_A = \sqrt{a_O^2 + (a_{A/O})_n^2 - 2a_O(a_{A/O})_n \cos\theta}$$

$$a_A = \sqrt{a_O^2 + (r\omega^2)^2 - 2a_O(r\omega^2)\cos\theta}$$

**Maple Worksheet**

```
> restart; with(plots):

> an:=r*omega^2;
            an := r ω²

> aA:=sqrt(a0^2+an^2-2*a0*an*cos(theta));
       aA := √(a0² + r² ω⁴ − 2 a0 r ω² cos(θ))

> r:=.8; a0:=3;
```

$r := 0.8$

$a0 := 3$

```
> omega:=2: aA1:=aA:
> omega:=4: aA2:=aA:
> omega:=6: aA3:=aA:
> p1:=plot([aA1,aA2,aA3], theta=0..2*Pi, color=black, labels=["theta (rads)",""],
title="acceleration of A (m/s^2)"):
> t1:=textplot([3,30,"6 rad/s"]):
> t2:=textplot([3,17,"4 rad/s"]):
> t3:=textplot([3,7.5,"2 rad/s"]):
> display(p1,t1,t2,t3);
```

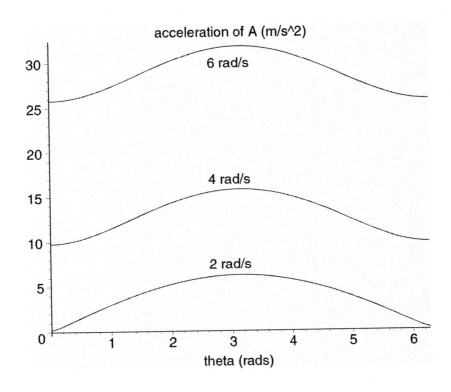

## 5.6 Sample Problem 5/15 (Absolute Motion)

The common configuration of a reciprocating engine is that of the slider crank mechanism shown. If crank $OB$ has a clockwise rotational speed of 1500 rev/min; (a) Plot $v_A$ and $v_G$ versus time for two revolutions of the crank. (b) Plot $a_A$ and $a_G$ versus time for two revolutions of the crank.

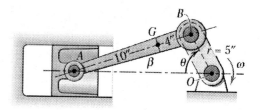

### Problem Formulation

This problem appears in sample problems 5/9 and 5/15 in your text. Sample problem 5/9 considers a relative velocity analysis while sample problem 5/15 uses a relative acceleration analysis. Generally speaking, the easiest approach to use with a computer is an absolute motion analysis, provided you have software capable of doing symbolic algebra and calculus such as Maple. We will use the present problem to illustrate this approach.

We start by using the law of sines ($\sin(\beta)/r = \sin(\theta)/l$) to express $\beta$ as a function of $\theta$

$$\beta = \sin^{-1}\left(\frac{r}{l}\sin\theta\right)$$

where $l$ is the length of connecting rod $AB$ and $\beta$ is the angle between $AB$ and the horizontal. Now place an $x$-$y$ coordinate system at $O$ with $x$ positive to the right and $y$ positive up and write expressions for the coordinates of $A$ and $G$ in terms of $\theta$ and $\beta$

$$x_A = -r\cos\theta - l\cos\beta$$

$$x_G = -r\cos\theta - \bar{r}\cos\beta \qquad y_G = (l - \bar{r})\sin\beta$$

where $\bar{r}$ is the distance from $B$ to $G$ (4 in. in the figure). All that is needed to find the velocities $v_A$, $v_{Gx}$, and $v_{Gy}$ is to differentiate these expressions with respect to time. The magnitude of the velocity of $G$ is then found from

$$v_G = \sqrt{v_{Gx}^2 + v_{Gy}^2}$$

The accelerations $a_A$, $a_{Gx}$, and $a_{Gy}$ are then found by differentiating $v_A$, $v_{Gx}$, and $v_{Gy}$ with the magnitude of the acceleration of $G$ being obtained from,

$$a_G = \sqrt{a_{Gx}^2 + a_{Gy}^2}$$

Since we will be differentiating with respect to time, the first thing we will do in the computer program is to define $\theta$ as a function of time. Then, when we write the above expressions for $\beta$, $x_A$, $x_G$, and $y_G$, the computer will automatically substitute for $\theta$ rendering each of these as functions of time. Assuming that $\theta$ is initially zero,

$$\theta(t) = \omega t = \frac{1500(2\pi)}{60} t = 157.1t$$

The problem statement asks us to plot versus time for two revolutions ($\theta = 4\pi$ radians) of the crank. The time required for two revolutions is $4\pi/157.1 = 0.08$ sec.

## Maple Worksheet

```
> restart; with(plots):
> theta:=omega*t;
```
$$\theta := \omega t$$

```
> beta:=arcsin(r/L*sin(theta));
```
$$\beta := \arcsin\left(\frac{r \sin(\omega t)}{L}\right)$$

```
> xA:=-r*cos(theta)-L*cos(beta);
```
$$xA := -r \cos(\omega t) - L \sqrt{1 - \frac{r^2 \sin(\omega t)^2}{L^2}}$$

```
> xG:=-r*cos(theta)-rb*cos(beta);
```
$$xG := -r \cos(\omega t) - rb \sqrt{1 - \frac{r^2 \sin(\omega t)^2}{L^2}}$$

```
> yG:=(L-rb)*sin(beta);
```
$$yG := \frac{(L - rb) r \sin(\omega t)}{L}$$

```
> vA:=diff(xA,t);
```
$$vA := r \sin(\omega t) \omega + \frac{r^2 \sin(\omega t) \cos(\omega t) \omega}{L \sqrt{1 - \frac{r^2 \sin(\omega t)^2}{L^2}}}$$

Note that, in what follows, the output for several results have been suppressed by ending the line with a colon ":". This is done to conserve space as the results are rather messy, especially the accelerations.

```
> vGx:=diff(xG,t):
> vGy:=diff(yG,t):
> vG:=sqrt(vGx^2+vGy^2);
```

$$vG := \sqrt{\left( r \sin(\omega t)\,\omega + \frac{rb\,r^2 \sin(\omega t)\cos(\omega t)\,\omega}{\sqrt{1 - \frac{r^2 \sin(\omega t)^2}{L^2}}\,L^2} \right)^2 + \frac{(L - rb)^2\,r^2 \cos(\omega t)^2\,\omega^2}{L^2}}$$

```
> aA:=diff(vA,t);
```

$$aA := r \cos(\omega t)\,\omega^2 + \frac{r^4 \sin(\omega t)^2 \cos(\omega t)^2\,\omega^2}{L^3 \left(1 - \frac{r^2 \sin(\omega t)^2}{L^2}\right)^{(3/2)}} + \frac{r^2 \cos(\omega t)^2\,\omega^2}{L \sqrt{1 - \frac{r^2 \sin(\omega t)^2}{L^2}}} - \frac{r^2 \sin(\omega t)^2\,\omega^2}{L \sqrt{1 - \frac{r^2 \sin(\omega t)^2}{L^2}}}$$

```
> aGx:=diff(vGx,t):
> aGy:=diff(vGy,t):
> aG:=sqrt(aGx^2+aGy^2);
```

$$aG := \text{sqrt}\Bigg( \left( r \cos(\omega t)\,\omega^2 + \frac{rb\,r^4 \sin(\omega t)^2 \cos(\omega t)^2\,\omega^2}{\left(1 - \frac{r^2 \sin(\omega t)^2}{L^2}\right)^{(3/2)} L^4} + \frac{rb\,r^2 \cos(\omega t)^2\,\omega^2}{\sqrt{1 - \frac{r^2 \sin(\omega t)^2}{L^2}}\,L^2} \right.$$

$$\left. - \frac{rb\,r^2 \sin(\omega t)^2\,\omega^2}{\sqrt{1 - \frac{r^2 \sin(\omega t)^2}{L^2}}\,L^2} \right)^2 + \frac{(L - rb)^2\,r^2 \sin(\omega t)^2\,\omega^4}{L^2} \Bigg)$$

```
>
> r:=5/12: L:=14/12: omega:=157.1: rb:=4/12:
> p1:=plot([vA,vG],t=0..0.08,labels=[`t (sec)`,` `], title="velocity (ft/s)"):
> t1:=textplot([.02,55,"vG"]): t2:=textplot([.02,20,"vA"]):
> display(p1, t1, t2);
```

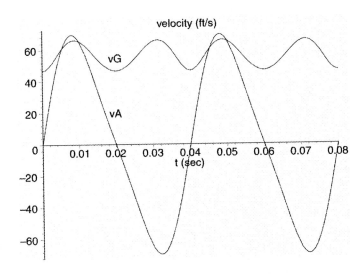

> p2:=plot([aA,aG],t=0..0.08,color=black,labels=[`t (sec)`,` `], title="acceleration (ft/s^2)"):
> t3:=textplot([.02,10000,"aG"]): t4:=textplot([.02,-5500,"aA"]):
> display(p2, t3, t4);

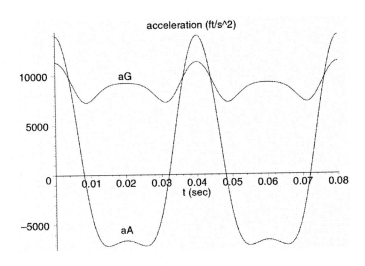

*For further study*

Suppose you wanted to solve this problem for the case where crank *OB* has a constant angular acceleration $\alpha = 60$ rad/s$^2$. It turns out that you can solve this problem using exactly the same approach as above, changing only one line defining the dependence of $\theta$ upon time. Assuming that the system starts from rest at $\theta = 0$, the appropriate expression for $\theta$ is $\theta(t) = \frac{1}{2}\alpha t^2 = 30t^2$. The accelerations for this case are shown below in case you want to give it a try.

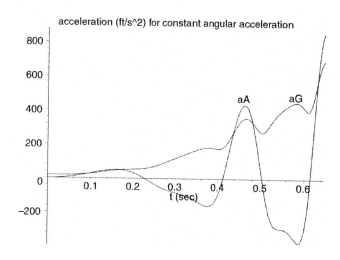

# PLANE KINETICS OF RIGID BODIES

# 6

This chapter concerns the motion (translation and rotation) of rigid bodies that results from the action of unbalanced external forces and moments. In problem 6.1, five equations are solved for five unknowns. The problem illustrates an alternative to the blunt (but straightforward) simultaneous solution of multiple equations. Instead, the equations are solved in such a way that there is never more than one unknown. In this way, the results are immediately obtained via automatic substitution in Maple, thus avoiding some tedious algebra. The force at the hinge of a pendulum is plotted versus the angular position of the pendulum in problem 6.2. The algebra is rather simple in this case and Maple is used primarily for purposes of plotting. Problems 6.3 and 6.4 consider rigid bodies in general plane motion. *solve* is used in problem 6.3 to solve two equations simultaneously for two unknowns. The maximum acceleration of a point on the rigid body is then obtained by using *diff* and *solve*. In this problem Maple is unable to find a symbolic result and automatically switches to a numerical approach finding five solutions to the equation. This problem is also interesting since a very natural "guess" of the value for the maximum acceleration turns out to be incorrect. Problem 6.4 is an example of a kinetics problem that also requires some kinematics. The angular acceleration of a bar is determined by summing moments. The kinematic equation $\omega d\omega = \alpha d\theta$ is then integrated to obtain the angular velocity. Problem 6.5 is an interesting work and energy problem that is complicated considerably by the fact that a spring is engaged for only part of the motion of a rotating bar. Symbolic algebra simplifies this problem considerably, though it is still rather tedious. It is common in Dynamics to find problems that require a combination of methods for their solution. Problem 6.6 is a good example involving both conservation of momentum and work/energy.

## 6.1 Sample Problem 6/2 (Translation)

The vertical bar *AB* has a mass of *m* =150 kg
with center of mass *G* midway between the ends.
The bar is elevated from rest at $\theta = 0$ by means
of the parallel links of negligible mass, with a
couple *M* applied to the lower link at *C*. Plot the
force at *A* and at *B* as functions of $\theta$ (between 0
and 60°) for two cases, (a) a constant couple *M*
= 5 kN-m, and (b) a constant angular
acceleration $\alpha = 5$ rad/sec$^2$.

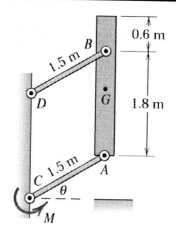

### Problem Formulation

The free-body diagram (FBD) and mass
acceleration diagram (MAD) are shown to
the right. Since the vertical bar undergoes
curvilinear translation, the acceleration of all
points on the bar will be identical. Thus, we
can obtain the acceleration of *G* immediately
from that of point *A* which moves in a
circular path about *C*. Also note that BD is a
two force member since the mass is
negligible.

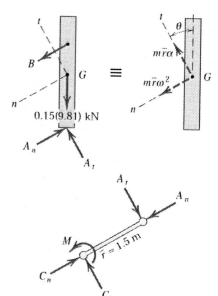

Here we take a somewhat different approach
than that in your textbook. The main reason
for this is that the case of constant moment
and constant angular acceleration require a
slightly different approach. The difference
between the two approaches is easiest seen
by writing all the equations first.

From the free-body diagram of the connecting link *AC*,

$$[\Sigma M_C = 0] \qquad M - \bar{r}A_t = 0 \qquad\qquad (1)$$

As pointed out in the sample problem, the force and moment equations are
identical to the equilibrium equations whenever the mass is negligible.

From the free-body diagram of the vertical bar,

$$\left[\Sigma M_A = m\bar{a}d\right] \qquad r_{AB}B\cos\theta = m\bar{r}\omega^2\cos\theta\; r_{AG} + m\bar{r}\alpha\sin\theta\; r_{AG} \qquad (2)$$

$$\left[\Sigma F_t = m\bar{a}_t\right] \qquad A_t - mg\cos\theta = m\bar{r}\alpha \qquad (3)$$

$$\left[\Sigma F_n = m\bar{a}_n\right] \qquad B - A_n + mg\sin\theta = m\bar{r}\omega^2 \qquad (4)$$

And, finally, the kinematics equation

$$\int\limits_0^\omega \omega\, d\omega = \int\limits_0^\theta \alpha\, d\theta \qquad (5)$$

At this point we have a total of 5 equations and 6 unknowns ($M$, $\alpha$, $\omega$, $A_t$, $A_n$, and $B$). In part (a), $M$ is specified while in part (b) $\alpha$ is given. In both cases we will have 5 equations and 5 unknowns; however, it will be necessary to solve no more than one equation at a time provided they are done in the right order. This is what yields a different procedure for parts (a) and (b).

### Part (a)

With $M$ known we can find $A_t$ from Equation (1) and then substitute the result into Equation (3) to get $\alpha$ as a function of $\theta$.

$$A_t = \frac{M}{\bar{r}}$$

$$\alpha = \frac{A_t}{m\bar{r}} - \frac{g}{\bar{r}}\cos\theta$$

As usual, notice that there is no need to make an explicit substitution of $A_t$ into $\alpha$. The computer will make such substitutions automatically. Now we substitute $\alpha$ into Equation (5) and integrate,

$$\frac{1}{2}\omega^2 = \int\limits_0^\theta \left(\frac{A_t}{m\bar{r}} - \frac{g}{\bar{r}}\cos\theta\right)d\theta$$

$$\omega^2 = 2\left(\frac{A_t}{m\bar{r}}\theta - \frac{g}{\bar{r}}\sin\theta\right)$$

Finally, we can substitute into Equations (2) and (4) to find $B$ and then $A_n$.

$$B = \frac{m\bar{r}\,r_{AG}}{r_{AB}\cos\theta}\left(\omega^2\cos\theta + \alpha\sin\theta\right)$$

$$A_n = B + mg\sin\theta - m\bar{r}\omega^2$$

$$A = \sqrt{A_n^2 + A_t^2}$$

### Part (b)

With $\alpha$ known instead of $M$ we have to take a different approach, starting with Equations (5) and (3),

$$\omega^2 = 2\int_0^\theta \alpha\, d\theta = 2\alpha\int_0^\theta d\theta = 2\alpha\theta$$

$$A_t = mg\cos\theta + m\bar{r}\alpha$$

Now we can substitute into Equations (1), (2) and (4) to find $M$, $B$ and then $A_n$.

$$M = \bar{r}A_t$$

$$B = \frac{m\bar{r}\,r_{AG}}{r_{AB}\cos\theta}\left(\omega^2\cos\theta + \alpha\sin\theta\right)$$

$$A_n = B + mg\sin\theta - m\bar{r}\omega^2$$

$$A = \sqrt{A_n^2 + A_t^2}$$

### Maple Worksheet

```
> restart;
part (a)
> A:=sqrt(An^2+At^2);
```
$$A := \sqrt{An^2 + At^2}$$

```
> An:=B+m*g*sin(theta)-m*r*omegasq;
```

$$An := B + m\, g\, \sin(\theta) - m\, r\, omegasq$$

> B:=m*r*AG/AB/cos(theta)*(omegasq*cos(theta)+alpha*sin(theta));

$$B := \frac{m\, r\, AG\, (omegasq\, \cos(\theta) + \alpha\, \sin(\theta))}{AB\, \cos(\theta)}$$

> alpha:=At/m/r-g/r*cos(theta);

$$\alpha := \frac{At}{m\, r} - \frac{g\, \cos(\theta)}{r}$$

> omegasq:=2*(At*theta/m/r-g/r*sin(theta));

$$omegasq := \frac{2\, At\, \theta}{m\, r} - \frac{2\, g\, \sin(\theta)}{r}$$

> At:=M/AC;

$$At := \frac{M}{AC}$$

> M:=5: m:=0.15: r:=1.5: AB:=1.8: AG:=1.2: AC:=r: g:=9.81:
> evalf(subs(theta=30*Pi/180,B)); # check the result obtained in the sample problem
   2.138606265

> plot([A,B],theta=0..60*Pi/180,color=black,labels=["theta (radians)","Force (kN)"],title="part a");

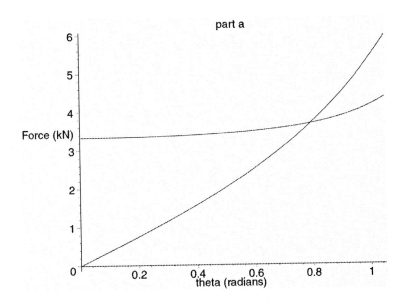

Part (b)
> restart;
> A:=sqrt(An^2+At^2);
$$A := \sqrt{An^2 + At^2}$$

> An:=B+m*g*sin(theta)-m*r*omegasq;
$$An := B + m\,g\sin(\theta) - m\,r\,omegasq$$

> B:=m*r*AG/AB/cos(theta)*(omegasq*cos(theta)+alpha*sin(theta));
$$B := \frac{m\,r\,AG\,(omegasq\cos(\theta) + \alpha\sin(\theta))}{AB\cos(\theta)}$$

> omegasq:=2*alpha*theta;
$$omegasq := 2\,\alpha\,\theta$$

> At:=m*g*cos(theta)+m*r*alpha;
$$At := m\,g\cos(\theta) + m\,r\,\alpha$$

> M:=AC*At;
$$M := AC\,(m\,g\cos(\theta) + m\,r\,\alpha)$$

> alpha:=5: m:=0.15: r:=1.5: AB:=1.8: AG:=1.2: AC:=r: g:=9.81:
> evalf(subs(theta=30*Pi/180,B)); # check the result obtained in the sample problem
            1.218410865

> plot([A,B],theta=0..60*Pi/180,color=black,labels=["theta (radians)","Force (kN)"],title="part b");

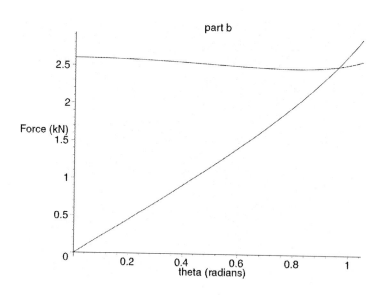

## 6.2 Sample Problem 6/4 (Fixed-Axis Rotation)

The pendulum has a mass of 7.5 kg with a mass center at $G$ and a radius of gyration about the pivot $O$ of 295 mm. If the pendulum is released from rest when $\theta = 0$, plot the total force supported by the bearing at $O$ along with its normal and tangential components as a function of $\theta$. Let $\theta$ range between 0 and 180°.

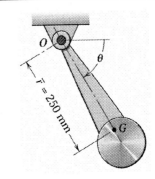

### Problem Formulation

The free body and mass acceleration diagrams are identical to those in the sample problem. The main difference in approach is that we will obtain results at an arbitrary angle $\theta$ rather than at 60°.

$$\left[\Sigma M_O = I_O \alpha\right] \qquad mg\bar{r}\cos\theta = mk_0^2\alpha \qquad \alpha = \frac{g\bar{r}}{k_0^2}\cos\theta$$

$$\left[\omega d\omega = \alpha d\theta\right] \qquad \int_0^\omega \omega d\omega = \frac{g\bar{r}}{k_0^2}\int_0^\theta \cos\theta d\theta = \frac{g\bar{r}}{k_0^2}\sin\theta$$

$$\omega^2 = \frac{2g\bar{r}}{k_0^2}\sin\theta$$

$$\left[\Sigma F_n = m\bar{r}\omega^2\right] \qquad O_n - mg\sin\theta = m\bar{r}\omega^2$$

$$\left[\Sigma F_t = m\bar{r}\alpha\right] \qquad -O_t + mg\cos\theta = m\bar{r}\alpha$$

After substituting for $\alpha$ and $\omega$ we have,

$$O_n = mg\left(1 + 2\frac{\bar{r}^2}{k_0^2}\right)\sin\theta \qquad\qquad O_t = mg\left(1 - \frac{\bar{r}^2}{k_0^2}\right)\cos\theta$$

The magnitude of the force at O is,

$$O = \sqrt{(O_n)^2 + (O_t)^2}$$

After substituting $\bar{r} = 0.25$ m, $k_0 = 0.295$ m, m = 7.5 kg, and g = 9.81 m/s$^2$ all forces will be functions of $\theta$ only.

### *Maple Worksheet*

> restart; with(plots):
> vB:=r*omega;
$$vB := r\,\omega$$

> On:=m*g*(1+2*rb^2/k0^2)*sin(theta);
$$On := m\,g\left(1 + \frac{2\,rb^2}{k0^2}\right)\sin(\theta)$$

> Ot:=m*g*(1-rb^2/k0^2)*cos(theta);
$$Ot := m\,g\left(1 - \frac{rb^2}{k0^2}\right)\cos(\theta)$$

> unprotect(O); # O is a protected name
> O:=sqrt(On^2+Ot^2);
$$O := \sqrt{m^2\,g^2\left(1 + \frac{2\,rb^2}{k0^2}\right)^2\sin(\theta)^2 + m^2\,g^2\left(1 - \frac{rb^2}{k0^2}\right)^2\cos(\theta)^2}$$

> rb:=0.25: k0:=0.295: m:=7.5: g:=9.81:
> p:=plot([On,Ot,O],theta=0..Pi,labels=["theta (rads)"," "],title="Force at O (N)"):
> t1:=textplot([.1,40,"O"]): t2:=textplot([.3,35,"On"]):
> t3:=textplot([.7,23,"Ot"]):
> display(p,t1,t2,t3);

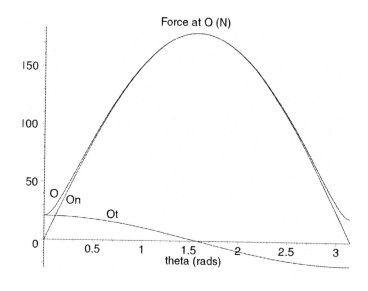

## 6.3 Problem 6/98 (General Plane Motion)

The slender rod of mass $m$ and length $l$ is released from rest in the vertical position with the small roller at end $A$ resting on the incline. (a) Determine the initial acceleration of $A$ ($a_A$) and plot $a_A$ versus $\theta$ for $0 \le \theta \le 90°$. (b) Determine the maximum value of $a_A$ over this range and the angle $\theta$ at which it occurs.

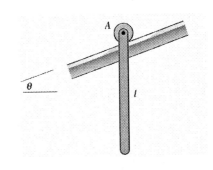

### Problem Formulation

The free-body and mass acceleration diagrams are shown to the right. Shown on the mass acceleration diagram are the two components of the acceleration of the center of mass $G$ obtained from the following kinematic relation,

*FBD*        *MAD*

$$\mathbf{a}_G = \mathbf{a}_A + (\mathbf{a}_{G/A})_n + (\mathbf{a}_{G/A})_t$$

where $(a_{G/A})_n = \dfrac{l}{2}\omega^2 = 0$, $(a_{G/A})_t = \dfrac{l}{2}\alpha$.

$$\left[\Sigma M_A = \bar{I}\alpha + m\bar{a}d\right] \qquad 0 = \frac{1}{12}ml^2\alpha + m\frac{l}{2}\alpha\frac{l}{2} - ma_A\frac{l}{2}\cos\theta$$

$$\left[\Sigma F_x = m\bar{a}_x\right] \qquad mg\sin\theta = m\left(a_A - \frac{l}{2}\alpha\cos\theta\right)$$

These two equations can be solved simultaneously to give,

$$\alpha = \frac{6(g/l)\sin\theta\cos\theta}{4 - 3\cos^2\theta} \qquad a_A = \frac{4g\sin\theta}{4 - 3\cos^2\theta}$$

At first, the answer to part (b) seems obvious. Intuitively, we would like to say that the maximum acceleration is $a_A = g$ and occurs at $\theta = 90°$. But this intuition neglects the effects of the bar's rotation upon the acceleration. As we will see below, the maximum acceleration is somewhat larger than $g$.

The maximum acceleration is obtained in the usual manner. The orientation where the maximum occurs is first found by solving the equation $da_A/d\theta = 0$

for $\theta$. This angle is then substituted back into the expression for $a_A$ to yield the maximum acceleration.

### Maple Worksheet

> restart;
> eqn1:=1/12*m*L^2*alpha+m*L/2*alpha*L/2-m*aA*L/2*cos(theta)=0;

$$eqn1 := \frac{1}{3} m L^2 \alpha - \frac{1}{2} m\, aA\, L \cos(\theta) = 0$$

> eqn2:=m*g*sin(theta)=m*(aA-L/2*alpha*cos(theta));

$$eqn2 := m\, g \sin(\theta) = m\left(aA - \frac{1}{2} L \alpha \cos(\theta)\right)$$

> soln:=solve({eqn1,eqn2},{aA,alpha});

$$soln := \{\, aA = -\frac{4\, g \sin(\theta)}{-4 + 3\cos(\theta)^2},\ \alpha = -\frac{6\, g \sin(\theta)\cos(\theta)}{L\,(-4 + 3\cos(\theta)^2)}\,\}$$

> assign(soln);
> g:=9.81:
> plot([theta*180/Pi,aA,theta=0..Pi/2],labels=["theta (degrees)","aA (m/s^2)"], labeldirections=[HORIZONTAL,VERTICAL]);

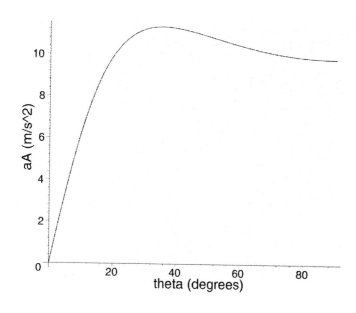

### Part (b)

> daA:=diff(aA,theta);

$$daA := -39.24 \frac{\cos(\theta)}{-4 + 3\cos(\theta)^2} - \frac{235.44\sin(\theta)^2\cos(\theta)}{(-4 + 3\cos(\theta)^2)^2}$$

> solve(daA=0,theta);
          1.570796327, 2.526112945, 0.6154797087

Maple has found three solutions, only two of which are in the range from 0 to 90°. These two solutions are $\pi/2$ (90°) and 0.6155 rad (35.3°). Referring to the figure above we see that the second solution corresponds to a maximum.

> evalf(subs(theta=.6154797087,aA));
          11.32761228

Thus, $(a_A)_{max} = 11.33$ m/s$^2$ when $\theta = 35.3°$.

---

## 6.4 Problem 6/104 (General Plane Motion)

The uniform 12-ft pole is hinged to the truck bed and released from the vertical position as the truck starts from rest with an acceleration $a$. If the acceleration remains constant during the motion of the pole, derive an expression for the angular velocity $\omega$ in terms of $a$, $g$, and $L$ where $L$ is the length of the pole. Plot $\omega$ versus $\theta$ for $a = 3$ ft/s$^2$.

### Problem Formulation

The free-body and mass acceleration diagrams are shown to the right. Shown on the mass acceleration diagram are the three components of the acceleration of the center of mass $G$ obtained from the following kinematic relation,

$$\mathbf{a}_G = \mathbf{a}_O + (\mathbf{a}_{G/O})_n + (\mathbf{a}_{G/O})_t$$

where $a_O = a$, $(a_{G/O})_n = \bar{r}\omega^2$, $(a_{G/O})_t = \bar{r}\alpha$, and $\bar{r} = L/2 = 6$ feet.

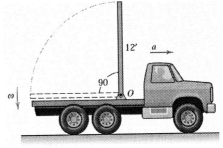

FBD          MAD

$$\left[\Sigma M_O = \bar{I}\alpha + m\bar{a}d\right] \qquad mg\frac{L}{2}\sin\theta = \frac{1}{12}mL^2\alpha + m\frac{L}{2}\alpha\frac{L}{2} - ma\frac{L}{2}\cos\theta$$

$$\alpha = \frac{3}{2L}\left(g\sin\theta + a\cos\theta\right)$$

Now we integrate the relation $\omega d\omega = \alpha d\theta$ to obtain,

$$\frac{1}{2}\omega^2 = \frac{3}{2L}\int_0^\theta \left(g\sin\theta + a\cos\theta\right)d\theta = \frac{3}{2L}\left(g\left(1-\cos\theta\right) + a\sin\theta\right)$$

$$\omega = \sqrt{\frac{3}{L}\left(g\left(1-\cos\theta\right) + a\sin\theta\right)}$$

## *Maple Worksheet*

```
> restart;
> eqn:=m*g*L/2*sin(theta)=1/12*m*L^2*alpha+m*(L/2)^2*alpha-m*a*L/2*cos(theta);
```
$$eqn := \frac{1}{2}\,m\,g\,L\,\sin(\theta) = \frac{1}{3}\,m\,L^2\,\alpha - \frac{1}{2}\,m\,a\,L\,\cos(\theta)$$

```
> alpha:=solve(eqn,alpha);
```
$$\alpha := \frac{3}{2}\,\frac{g\,\sin(\theta) + a\,\cos(\theta)}{L}$$

```
> I1:=int(subs(theta=x,alpha),x=0..theta);
```
$$I1 := -\frac{3}{2}\,\frac{g\,\cos(\theta) - a\,\sin(\theta) - g}{L}$$

```
> omega:=sqrt(2*I1);
```
$$\omega := \sqrt{-\frac{3\,g\,\cos(\theta) - 3\,a\,\sin(\theta) - 3\,g}{L}}$$

```
> L:=12: g:=32.2: a:=3:
> plot([theta*180/Pi,omega,theta=0..Pi/2],labels=["theta (deg)","omega
(rad/s)"],labeldirections=[HORIZONTAL,VERTICAL]);
```

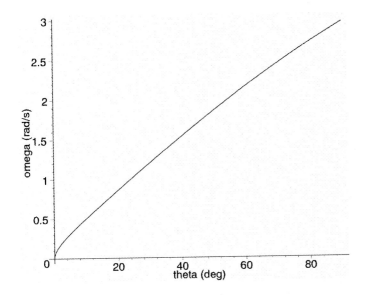

## 6.5 Sample Problem 6/10 (Work and Energy)

The 4-ft slender bar weighs 40 lb with a mass center at $B$ and is released from rest in the position for which $\theta$ is essentially zero. Point $B$ is confined to move in the smooth vertical guide, while end $A$ moves in the smooth horizontal guide and compresses the spring as the bar falls. Plot the angular velocity of the bar and the velocities of $A$ and $B$ as a function of $\theta$ from 0 to 90°. The stiffness of the spring is 30 lb/in.

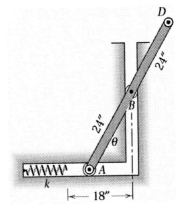

### Problem Formulation

From the figure to the right, the lengths $CB$ and $CA$ are $2\sin\theta$ and $2\cos\theta$ respectively. Using the instantaneous center $C$ we can write the two velocities in terms of the angular velocity $\omega$.

$$v_A = CA\omega = 2\omega\cos\theta \qquad v_B = CB\omega = 2\omega\sin\theta$$

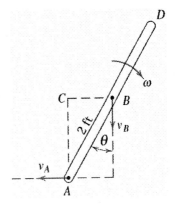

Now we need to divide the range for $\theta$ into two distinct intervals depending upon whether or not the spring has been engaged. Since the velocities for $A$ and $B$ are known in terms of $\omega$ and $\theta$, we need to find only the angular velocity in these two intervals. From the diagram we see that $A$ will first contact the spring at an angle $\theta = \sin^{-1}(18/24) = 0.8481$ rads (48.6°).

**(a)** Before the spring is engaged $(\theta \leq 48.6°)$.

$$[T = \frac{1}{2}m\bar{v}^2 + \frac{1}{2}\bar{I}\omega^2] \quad \Delta T = \frac{1}{2}\frac{40}{32.2}(2\omega\cos\theta)^2 + \frac{1}{2}\left(\frac{1}{12}\frac{40}{32.2}4^2\right)\omega^2$$

$$\Delta T = 0.8282\left(4 - 3\cos^2\theta\right)\omega^2$$

$$[\Delta V_g = W\Delta h] \qquad \Delta V_g = 40(2\cos\theta - 2) = 80(\cos\theta - 1)$$

We now substitute into the energy equation $U'_{1-2} = \Delta T + \Delta V_g = 0$,

$0 = 0.8282\left(4 - 3\cos^2\theta\right)\omega^2 + 80(\cos\theta - 1)$, from which we find

$$\omega = 9.829\sqrt{\frac{1 - \cos\theta}{4 - 3\cos^2\theta}}$$

**(b)** After the spring is engaged $(48.6 \leq \theta \leq 90°)$. The kinetic and potential energies are the same as in part (a). At any angle $\theta$, point $A$ has moved $2\sin\theta$ feet to the left. Thus, the spring is compressed by $2\sin\theta - 18/12$ feet.

$$[V_e = \frac{1}{2}kx^2] \qquad \Delta V_e = \frac{1}{2}\left(30\frac{lb}{in}\right)\left(12\frac{in}{ft}\right)\left(2\sin\theta - \frac{18}{12}\right)^2 - 0$$

$$\Delta V_e = 180\left(2\sin\theta - \frac{3}{2}\right)^2$$

Again, we substitute into the energy equation $U'_{1-2} = \Delta T + \Delta V_g + \Delta V_e = 0$,

$$0 = 0.8282\left(4 - 3\cos^2\theta\right)\omega^2 + 80(\cos\theta - 1) + 180\left(2\sin\theta - \frac{3}{2}\right)^2$$

$$\omega = 2.457 \sqrt{\frac{216\sin\theta - 16\cos\theta - 144\sin^2\theta - 65}{4 - 3\cos^2\theta}}$$

**Maple Worksheet**

> restart; with(plots):

> theta0:=arcsin(18/24);

$$\theta0 := \arcsin\left(\frac{3}{4}\right)$$

In the following, subscript a is for part (a) where $\theta$ is between 0 and $\theta_0$ (48.6°) while subscript b is for part (b) where $\theta$ is greater than $\theta_0$.

> DT:=0.8282*(4-3*cos(theta)^2)*omega^2;

$$DT := .8282\,(4 - 3\cos(\theta)^2)\,\omega^2$$

> DVg:=80*(cos(theta)-1);

$$DVg := 80\cos(\theta) - 80$$

> DVe:=180*(2*sin(theta)-3/2)^2;

$$DVe := 180\left(2\sin(\theta) - \frac{3}{2}\right)^2$$

> U12[a]:=0=DT+DVg;

$$U12_a := 0 = .8282\,(4 - 3\cos(\theta)^2)\,\omega^2 + 80\cos(\theta) - 80$$

> U12[b]:=DT+DVg+DVe;

$$U12_b := .8282\,(4 - 3\cos(\theta)^2)\,\omega^2 + 80\cos(\theta) - 80 + 180\left(2\sin(\theta) - \frac{3}{2}\right)^2$$

> solve(U12[a],omega);

$$\frac{9.828276825\,\sqrt{(-4. + 3.\cos(\theta)^2)(-1. + \cos(\theta))}}{-4. + 3.\cos(\theta)^2},$$

$$-\frac{9.828276825\,\sqrt{(-4. + 3.\cos(\theta)^2)(-1. + \cos(\theta))}}{-4. + 3.\cos(\theta)^2}$$

It shouldn't be surprising that Maple has found two solutions to the equation. As you can see, they have the same magnitude with one being positive and the other negative. The solution of interest is the positive one which turns out to be the second solution.

> omega[a]:=%[2]; # assigns the second solution to $\omega_a$

$$\omega_a := -200 \frac{\sqrt{41410}\sqrt{(-4+3\cos(\theta)^2)(-1+\cos(\theta))}}{-16564+12423\cos(\theta)^2}$$

> solve(U12[b],omega);

Note that no output followed this statement which means that Maple was unable to find the solution to this equation for some reason. This is somewhat odd in that some earlier versions of Maple are able to solve the equation with no difficulty. Another oddity is that the equation would be rather easy to solve by hand. Rather than resort to solving by hand let's try a little trick and have Maple solve for omega squared.

> solve(U12[b],omega^2);
```
Warning, solving for expressions other than names or functions is not
recommended.
```

$$\frac{6.037189085\,(16.\cos(\theta)+65.+144.\sin(\theta)^2-216.\sin(\theta))}{-4.+3.\cos(\theta)^2}$$

We were successful in this case but you might want to notice the warning.

> omega[b]:=sqrt(%);

$$\omega_b := \sqrt{\frac{96.59502536\cos(\theta)+392.4172905+869.3552282\sin(\theta)^2-1304.032842\sin(\theta)}{-4.+3.\cos(\theta)^2}}$$

We need to be careful to plot the two results only over the range for which they are valid. (a) for $\theta$ between 0 and $\theta_0$ (48.6°) and (b) for $\theta$ greater than $\theta_0$. Also note that we use a parametric plot in order to plot versus $\theta$ in degrees instead of radians.

> pw1:=plot([theta*180/Pi,omega[a],theta=0..theta0], labels=["theta (deg)","angular velocity (rad/s)"],labeldirections= [HORIZONTAL,VERTICAL]):
> pw2:=plot([theta*180/Pi,omega[b],theta=theta0..Pi/2]):

> display(pw1,pw2);

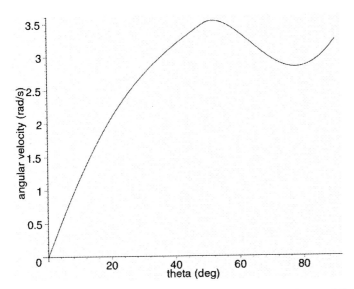

Now we find the velocities. Remember that capital A and B refer to points A and B while lower case a and b refer to cases (a) and (b).

> vA[a]:=2*omega[a]*cos(theta):
> vA[b]:=2*omega[b]*cos(theta):
> vB[a]:=2*omega[a]*sin(theta):
> vB[b]:=2*omega[b]*sin(theta):

> pA1:=plot([theta*180/Pi,vA[a],theta=0..theta0], labels=["theta (deg)","velocity (ft/s)"],labeldirections=[HORIZONTAL,VERTICAL]):
> pA2:=plot([theta*180/Pi,vA[b],theta=theta0..Pi/2]):
> pB1:=plot([theta*180/Pi,vB[a],theta=0..theta0]):
> pB2:=plot([theta*180/Pi,vB[b],theta=theta0..Pi/2]):
> t1:=textplot([60,3.8,"vA"]): t2:=textplot([60,6.1,"vB"]):

> display(pA1,pA2,pB1,pB2,t1,t2);

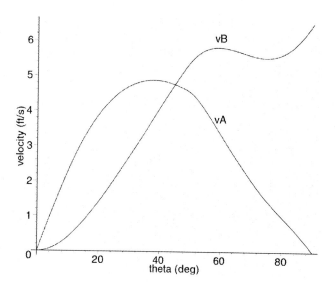

*For Further Study*

A striking feature of the velocity curves above is the sudden change in shape when the spring is engaged at about $\theta = 49°$. The details depend very much upon the relative magnitudes of the weight and spring constant. To illustrate, the figure below shows the angular velocity for several different values of the spring constant $k$.

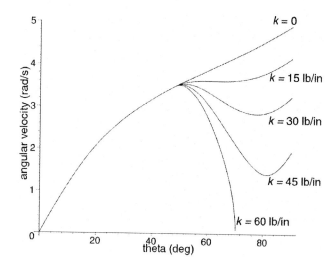

Note that for stiff springs, the angular velocity goes to zero before reaching $\theta = 90°$. The physical explanation for this is that, for a stiff spring, the bar will rebound before it reaches the horizontal position.

## 6.6 Problem 6/206 (Impulse/Momentum)

Determine the minimum velocity $v$ that the wheel may have and just roll over the obstruction. The centroidal radius of gyration of the wheel is $k$, and it is assumed that the wheel does not slip. Plot $v$ versus $h$ for three cases: $k = \frac{1}{2}, \frac{3}{4}$, and 1 m. For each case take $r = 1$ m.

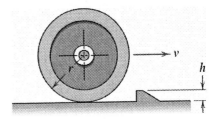

### Problem Formulation

*During Impact*: Conservation of Angular Momentum

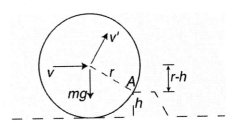

As usual, we neglect the angular impulse of the weight during the short interval of impact. With this assumption we have conservation of angular momentum about point $A$. Immediately before impact, the center of the wheel is not moving in a circular path about $A$ and we need to use the formula for general plane motion. Note that $\bar{I} = mk^2$.

$$H_A = \bar{I}\omega + m\bar{v}d = mk^2\omega + mv(r-h) = mk^2\frac{v}{r} + mv(r-h)$$

We will use primes to denote the state immediately after impact. Since the wheel now rotates about $A$ we can use the simpler formula $H_A = I_A\omega$. Note that, by the parallel axis theorem, $I_A = \bar{I} + mr^2 = m(k^2 + r^2)$.

$$H_A' = I_A\omega' = (k^2 + r^2)\frac{v'}{r}$$

Setting $H_A = H_A'$ and solving yields,

$$v' = v\left(1 - \frac{rh}{k^2 + r^2}\right)$$

*After Impact*: Work-Energy

$$\Delta T + \Delta V_g = 0 = \frac{1}{2}I_A\left(0^2 - \omega'^2\right) + mgh$$

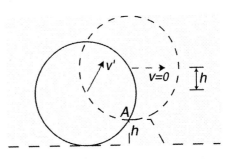

$$\frac{1}{2}m\left(k^2 + r^2\right)\left(\frac{v'}{r}\right)^2 = mgh$$

Substituting the result for $v'$ into the above equation followed by simplification yields,

$$v = \frac{r\sqrt{2gh\left(k^2 + r^2\right)}}{k^2 + r^2 - rh}$$

**Maple Worksheet**

```
> restart; with(plots):
> Ib:=m*k^2; IA:=m*(k^2+r^2);
```
$$Ib := m\,k^2$$

$$IA := m\,(k^2 + r^2)$$

```
> HA:=Ib*v/r+m*v*(r-h);
```
$$HA := \frac{m\,k^2\,v}{r} + m\,v\,(r - h)$$

```
> HAp:=IA*vp/r;
```
$$HAp := \frac{m\,(k^2 + r^2)\,vp}{r}$$

```
> vp:=solve(HA=HAp,vp);   # conservation of ang. mom.
```
$$vp := \frac{v\,(k^2 + r^2 - r\,h)}{k^2 + r^2}$$

```
> eqn:=1/2*IA*vp^2/r^2=m*g*h;   # work/energy
```
$$eqn := \frac{1}{2}\frac{m\,v^2\,(k^2 + r^2 - r\,h)^2}{(k^2 + r^2)\,r^2} = m\,g\,h$$

```
> solve(eqn,v);
```
$$\frac{\sqrt{2\,g\,h\,r^2 + 2\,g\,h\,k^2}\;r}{k^2 + r^2 - r\,h}, \quad -\frac{\sqrt{2\,g\,h\,r^2 + 2\,g\,h\,k^2}\;r}{k^2 + r^2 - r\,h}$$

Note that it is possible to copy Maple output and then paste it into an input line. This feature can save a lot of time typing. Here we copy the first solution above and then copy it into the following expression defining the velocity.

```
> v:=sqrt(2*g*h*r^2+2*g*h*k^2)*r/(k^2+r^2-r*h);
```
$$v := \frac{\sqrt{2\,g\,h\,r^2 + 2\,g\,h\,k^2}\;r}{k^2 + r^2 - r\,h}$$

```
> r:=1: g:=9.81:
> v1:=subs(k=1/2,v): v2:=subs(k=3/4,v): v3:=subs(k=1,v):
> p:=plot([v1,v2,v3],h=0..1,labels=["h (m)",""],title="velocity (m/s)"):
> t1:=textplot([.9,17,"k=1/2"]):t2:=textplot([.89,9.2,"k=3/4"]):
> t3:=textplot([.87,4.4,"k=1"]):

> display(p,t1,t2,t3);
```

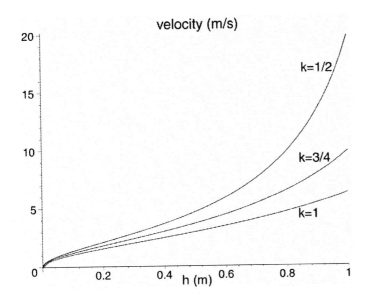

# INTRODUCTION TO THREE-DIMENSIONAL DYNAMICS OF RIGID BODIES

# 7

This chapter presents a brief introduction to rigid body dynamics in three dimensions. In problem 7.1, the general 3-D motion of three connected bars is investigated. In particular, the angular velocities of two of the bars are plotted versus the length of the third bar. *solve* is used to solve four equations symbolically for four unknowns. In problem 7.2 we consider a bent plate rotating about a fixed axis. The problem illustrates a simplified version of what engineers might do in a real design situation. Two dimensions of the bent plate are left as variables and the objective of the problem is to find all suitable values of those dimensions which satisfy several constraints simultaneously. Here we illustrate a graphical approach to this type of design problem.

## 7.1 Sample Problem 7/3 (General Motion)

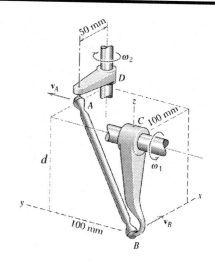

Crank $CB$ rotates about the horizontal axis with an angular velocity $\omega_1 = 6$ rad/s, which is constant for a short interval of motion that includes the position shown. Link $AB$ has a ball-and-socket fitting on each end and connects crank $DA$ with $CB$. Let the length of crank $CB$ be $d$ mm (instead of 100 mm as in the sample problem in your text) and plot $\omega_2$ and $\omega_n$ as a function of $d$ for $0 \le d \le 200$ mm. $\omega_2$ is the angular velocity of crank $DA$ while $\omega_n$ is the angular velocity of link $AB$.

### Problem Formulation

Our analysis will follow closely that in the sample problem in your text.

$$\mathbf{v}_A = \mathbf{v}_B + \boldsymbol{\omega}_n \times \mathbf{r}_{AB}$$

where   $\mathbf{v}_A = 50\omega_2\mathbf{j}$      $\mathbf{v}_B = 6d\mathbf{i}$      $\mathbf{r}_{AB} = 50\mathbf{i} + 100\mathbf{j} + d\mathbf{k}$

Substitution into the velocity equation gives

$$50\omega_2\mathbf{j} = 6d\mathbf{i} + \begin{vmatrix} \mathbf{i} & \mathbf{j} & \mathbf{k} \\ \omega_{nx} & \omega_{ny} & \omega_{nz} \\ 50 & 100 & d \end{vmatrix}$$

Expanding the determinant and equating the $\mathbf{i}$, $\mathbf{j}$, and $\mathbf{k}$ components yields the following three equations

$$d(6 + \omega_{ny}) - 100\omega_{nz} = 0 \qquad 50(\omega_2 - \omega_{nz}) + d\omega_{nx} = 0 \qquad 2\omega_{nx} - \omega_{ny} = 0$$

At this point we have three equations with four unknowns. As explained in the sample problem in your text, the fourth equation comes by requiring $\boldsymbol{\omega}_n$ to be normal to $\mathbf{v}_{AB}$

$$\boldsymbol{\omega}_n \bullet \mathbf{r}_{AB} = 50\omega_{nx} + 100\omega_{ny} + d\omega_{nz} = 0$$

These four equations will be solved simultaneously for $\omega_2$, $\omega_{nx}$, $\omega_{ny}$, and $\omega_{nz}$. Once this is done,

$$\omega_n = \sqrt{\omega_{nx}^2 + \omega_{ny}^2 + \omega_{nz}^2}$$

### Maple Worksheet

> restart; with(plots):
> eqn1:=d*(6+omega[ny])-100*omega[nz]=0;

$$eqn1 := d\,(6 + \omega_{ny}) - 100\,\omega_{nz} = 0$$

> eqn2:=50*(omega[2]-omega[nz])+d*omega[nx]=0;

$$eqn2 := 50\,\omega_2 - 50\,\omega_{nz} + d\,\omega_{nx} = 0$$

> eqn3:=2*omega[nx]-omega[ny]=0;

$$eqn3 := 2\,\omega_{nx} - \omega_{ny} = 0$$

> eqn4:=50*omega[nx]+100*omega[ny]+d*omega[nz]=0;

$$eqn4 := 50\,\omega_{nx} + 100\,\omega_{ny} + d\,\omega_{nz} = 0$$

> soln:=solve({eqn1,eqn2,eqn3,eqn4},{omega[2],omega[nx],omega[ny],omega[nz]});

$$soln := \{\,\omega_2 = \frac{3}{50}\,d,\ \omega_{nx} = -3\,\frac{d^2}{d^2 + 12500},\ \omega_{nz} = 750\,\frac{d}{d^2 + 12500},\ \omega_{ny} = -6\,\frac{d^2}{d^2 + 12500}\,\}$$

> assign(soln);
> omega[n]:=simplify(sqrt(omega[nx]^2+omega[ny]^2+omega[nz]^2));

$$\omega_n := 3\,\sqrt{5}\,\sqrt{\frac{d^2}{d^2 + 12500}}$$

> p:=plot([omega[2],omega[n]],d=0..200,labels=["d (mm)",""],title="angular velocity (rad/s)"):
> t1:=textplot([160,11,"omega_2"]):
> t2:=textplot([160,6.1,"omega_n"]):
> display(p,t1,t2);

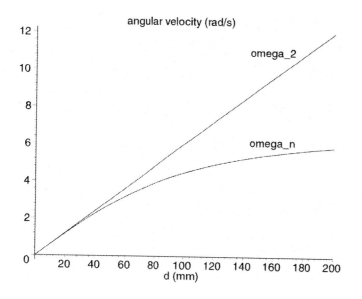

## 7.2 Sample Problem 7/6 (Kinetic Energy)

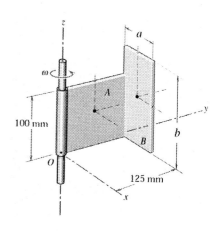

The bent plate has a mass of 70 kg per square meter of surface area and revolves around the $z$-axis at the rate $\omega =$ 30 rad/s. Let the dimensions of part $B$ be $a$ and $b$ where $a$ is the dimension parallel to the $x$-axis and $b$ is the dimension parallel to the $z$-axis. Part $A$ remains unchanged. (a) Find all suitable values for $a$ and $b$ which satisfy the following conditions: $a \leq 0.2$ m, $b \leq 0.6$ m, and $15 \leq T \leq 30$ J where $T$ is the kinetic energy of the plate. (b) Find $a$ and $b$ for the case where $T = 40$ J and $H_0 = 5$ N•m•s where $H_0$ is the magnitude of the angular momentum about $O$.

### Problem Formulation

Substitution of $\omega_x = 0$, $\omega_y = 0$, and $\omega_z = \omega$ into equations 7/11 and 7/18 of your text yields.

$$\mathbf{H}_O = \omega(-I_{xz}\mathbf{i} - I_{yz}\mathbf{j} + I_{zz}\mathbf{k})$$

$$H_O = \sqrt{H_{Ox}^2 + H_{Oy}^2 + H_{Oz}^2} = \omega\sqrt{I_{xz}^2 + I_{yz}^2 + I_{zz}^2}$$

$$T = \frac{1}{2} I_{zz} \omega^2$$

The moments and products of inertia for part $A$ remain unchanged. For part $B$, $m_B$ = 70$ab$ where $a$ and $b$ are in meters. The moments and products of inertia for part $B$ are

$$I_{zz} = \bar{I}_{zz} + md^2 = \frac{m_B}{12} a^2 + m_B \left[ (.125)^2 + \left( \frac{a}{2} \right)^2 \right]$$

$$I_{xz} = \bar{I}_{xz} + md_x d_z = 0 + m_B \left( \frac{a}{2} \right) \left( \frac{b}{2} \right)$$

$$I_{yz} = \bar{I}_{yz} + md_y d_z = 0 + m_B (0.125) \left( \frac{b}{2} \right)$$

The total moment and products of inertia are found by adding the above to those found for part $A$ (see the sample problem in your text). After substituting for $m_B$ and simplifying we have,

$$I_{zz} = 0.00456 + 1.094ab + 23.33a^3b$$
$$I_{xz} = 17.5a^2b^2 \qquad I_{yz} = 0.00273 + 4.375ab^2$$

Substitution of these results will give $H_O$ and $T$ as functions of $a$ and $b$.

*Part (a)*

The most efficient way to show the acceptable ranges for $a$ and $b$ is to find the required relationship between these two dimensions in order to satisfy the upper and lower bounds on $T$. This is accomplished by substituting these bounds for $T$ in the equation above and then solving that equation for $b$ as a function of $a$. To illustrate, consider the lower limit on $T$ (15 J). Substituting $T$ = 15 into the equation above gives

$$T = 15 = \frac{1}{2} I_{zz} (30)^2 = 2.052 + 492.2ab + 10,500a^3b$$

Solving for b, $\qquad b = \dfrac{0.078921}{a\left(3 + 64a^2\right)}$ $\qquad$ (for $T$ = 15 J)

A similar result can be obtained for the upper limit,

$$b = \frac{0.17035}{a\left(3 + 64a^2\right)} \qquad \text{(for } T = 30 \text{ J)}$$

Plotting these two functions defines the acceptable regions for $a$ and $b$.

*Part (b)*

This is similar to (a) except that we solve two equations ($T = 40$ and $H_0 = 5$) simultaneously for two unknowns, $a$ and $b$. The result is $a = 0.1211$ m and $b = 0.4852$ m.

### Maple Worksheet

> restart; with(plots):
> Izz:=m[B]/12*a^2+m[B]*(0.125^2+(a/2)^2)+0.00456;

$$Izz := \frac{1}{12} m_B a^2 + m_B\left(.015625 + \frac{1}{4} a^2\right) + .00456$$

> Ixz:=m[B]*a/2*b/2;

$$Ixz := \frac{1}{4} m_B a b$$

> Iyz:=m[B]*0.125*b/2+0.00273;

$$Iyz := .06250000000 m_B b + .00273$$

> H0:=omega*sqrt(Ixz^2+Iyz^2+Izz^2): # output has been suppressed
> T:=1/2*Izz*omega^2:

*Part (a)*
> omega:=30: m[B]:=70*a*b:
> b15:=solve(T=15, b);

$$b15 := .07892114286 \frac{1}{a\left(64. a^2 + 3.\right)}$$

> b30:=solve(T=30, b);

$$b30 := .1703497143 \frac{1}{a\left(64. a^2 + 3.\right)}$$

> p:=plot([b15,b30],a=0..0.2,y=0..0.6,labels=["a (m)","b (m)"],
labeldirections=[HORIZONTAL,VERTICAL]):
> t1:=textplot([.055,.6,"T = 15"]): t2:=textplot([.097,.6,"T = 30"]):
> display(p,t1,t2);

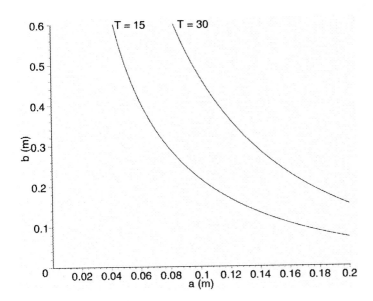

The two curves above represent the values of $a$ and $b$ for which $T$ is exactly 15 or 30 J. Thus, the acceptable values of $a$ and $b$ satisfying the condition $15 \leq T \leq 30$ J are all those combinations lying on or between the two curves.

*Part (b)*
```
> solve({H0=5,T=40},{a,b});
```
$\{b = -.4995916826, a = -.1186729553\}$,

$\quad\{b = .2865196782 + .2810897606I, a = -.09249618254 - .2040591236I\}$,

$\quad\{a = -.09249618254 + .2040591236I, b = .2865196782 - .2810897606I\}$,

$\quad\{b = .06437577589 - .7784870381I, a = -.002575144973 - .2549173146I\}$,

$\quad\{a = -.002575144973 + .2549173146I, b = .06437577589 + .7784870381I\}$,

$\quad\{b = -.05367082782 - .5793986603I, a = .003496405433 - .2650400901I\}$,

$\quad\{a = .003496405433 + .2650400901I, b = -.05367082782 + .5793986603I\}$,

$\quad\{a = .09037983691 - .2043122065I, b = -.2981058787 + .2845590159I\}$,

$\quad\{a = .09037983691 + .2043122065I, b = -.2981058787 - .2845590159I\}$,

$\quad\{b = .4851675635, a = .1210631257\}$

We see from the above that Maple has found ten solutions to our equations. Of these only the first and last are real. The first solution has negative values for $a$ and $b$ and thus can be excluded. This leaves only the last solution,

$$a = 0.1211 \text{ m}; \quad b = 0.4852 \text{ m}$$

# VIBRATION AND TIME RESPONSE

# 8

This chapter considers an important class of Dynamics problems that involve linear or angular oscillations of a body or structure about some equilibrium position or configuration. Very few of the homework problems in your text require you to actually plot the oscillations of a body versus time. For this reason, all of the problems in this chapter will involve such plots. This is very useful in visualizing the time response of a vibrating system, especially for the case of damped or forced vibrations. Problem 8.1 looks at the effects of damping coefficient upon time response while Problem 8.3 considers the effects of initial conditions.

## 8.1 Sample Problem 8/2 (Free Vibration of Particles)

The 8-kg body is moved 0.2 m to the right of the equilibrium position and released from rest at time t = 0. Plot the displacement as a function of time for three cases, c = 8, 32, and 56 N•s/m. The spring stiffness k is 32 N/m.

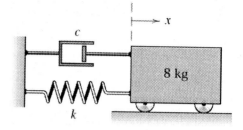

### Problem Formulation

As in the sample problem in your text, the natural circular frequency is $\omega_n = \sqrt{k/m} = \sqrt{32/8} = 2$ rad/sec. Now we find the damping ratio for each case (from $\zeta = c/2m\omega_n$) with the result $\zeta = 0.25$, 1, and 1.75 for c = 8, 32, and 56 N•s/m respectively.

**(a) $\zeta = 0.25$.** Since $\zeta < 1$, the system is underdamped. The damped natural frequency is $\omega_d = \omega_n\sqrt{1-\zeta^2} = 1.937$ rad/sec. The displacement and velocity are

$$x = Ce^{-\zeta\omega_n t}\sin(\omega_d t + \psi) = Ce^{-t/2}\sin(1.937t + \psi)$$

$$\dot{x} = -0.5Ce^{-t/2}\sin(1.937t + \psi) + 1.937Ce^{-t/2}\cos(1.937t + \psi)$$

From the initial conditions $x_0 = 0.2$ and $\dot{x}_0 = 0$ we find C = 0.207 m and $\psi = 1.318$ rad.

**(b) $\zeta = 1$.** For $\zeta = 1$, the system is critically damped. The displacement and velocity are

$$x = (A_1 + A_2 t)e^{-\omega_n t} = (A_1 + A_2 t)e^{-2t}$$

$$\dot{x} = A_2 e^{-2t} - 2(A_1 + A_2 t)e^{-2t}$$

From the initial conditions $x_0 = 0.2$ and $\dot{x}_0 = 0$ we find $A_1 = 0.2$ m and $A_2 = 0.4$ m/s.

**(c) $\zeta = 1.75$.** Since $\zeta > 1$, the system is overdamped. The displacement and velocity are

$$x = B_1 e^{\left(-\zeta+\sqrt{\zeta^2-1}\right)\omega_n t} + B_2 e^{\left(-\zeta-\sqrt{\zeta^2-1}\right)\omega_n t} = B_1 e^{-0.628t} + B_2 e^{-6.372t}$$

$$\dot{x} = -0.628 B_1 e^{-0.628t} - 6.372 B_2 e^{-6.372t}$$

From the initial conditions $x_0 = 0.2$ and $\dot{x}_0 = 0$ we find $B_1 = 0.222$ m and $B_2 = -0.0219$ m/s.

### *Maple Worksheet*

```
> restart; with(plots):
> x[a]:=C*exp(-t/2)*sin(1.937*t+psi);
```
$$x_a := C\, e^{(-1/2\,t)} \sin(1.937\,t + \psi)$$

```
> C:=0.207; psi:=1.318;
```
$$C := .207$$

$$\psi := 1.318$$

```
> x[b]:=(A1+A2*t)*exp(-2*t);
```
$$x_b := (A1 + A2\,t)\, e^{(-2\,t)}$$

```
> A1:=0.2; A2:=0.4;
```
$$A1 := .2$$

$$A2 := .4$$

```
> x[c]:=B1*exp(-0.628*t)+B2*exp(-6.372*t);
```
$$x_c := B1\, e^{(-.628\,t)} + B2\, e^{(-6.372\,t)}$$

```
> B1:=0.222; B2:=-0.0219;
```
$$B1 := .222$$

$$B2 := -.0219$$

```
> p:=plot([x[a],x[b],x[c]],t=0..5,color=black,labels=["time (sec)","x (m)"]):
> ta:=textplot([2.2,-0.08,"c = 8"]):
> tb:=textplot([1.6,0.06,"c = 32"]):
> tc:=textplot([1.6,.11,"c = 56"]):
> display(p,ta,tb,tc);
```

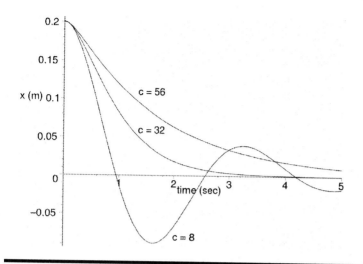

## 8.2 Problem 8/139 (Damped Free Vibration)

The mass of a critically damped system having a natural frequency $\omega_n$ is released from rest at an initial displacement $x_0$. (a) Determine the time $t$ required for the mass to reach the position $x = 0.1x_0$ if $\omega_n = 4$ rad/s. (b) Plot the non-dimensional displacement $x/x_0$ for $\omega_n = 2$, 4, and 8 rad/s.

### Problem Formulation

Start with the equation for the critically damped case on page 606 of your text.

$$x = \left(A_1 + A_2 t\right)e^{-\omega_n t}$$

First we have to evaluate the constants in terms of $x_0$ using the initial conditions. This will require the derivative of $x$.

$$\dot{x} = A_2 e^{-\omega_n t} - \left(A_1 + A_2 t\right)\omega_n e^{-\omega_n t}$$

Now, substituting the initial conditions,

$$x(t = 0) = x_0 = A_1$$

$$\dot{x}(t=0) = 0 = A_2 - A_1\omega_n$$

These two equations give

$$A_1 = x_0 \text{ and } A_2 = x_0\omega_n$$

$$x = (x_0 + x_0\omega_n t)e^{-\omega_n t}$$

or $$\eta = (1 + \omega_n t)e^{-\omega_n t}$$

where $\eta$ is the non-dimensional displacement $x/x_0$.

Part (a) will require the solution to the equation

$$0.1 = (1 + 4t)e^{-4t}$$

For part (b) we will simply plot $\eta$ for three different natural frequencies $\omega_n$.

## Maple Worksheet

> restart; with(plots):
> eta:=(1+omega[n]*t)*exp(-omega[n]*t);
$$\eta := (1 + \omega_n t)e^{(-\omega_n t)}$$

> solve(eta=1/10,t);
$$-\frac{\text{LambertW}\left(-\frac{1}{10}e^{(-1)}\right)+1}{\omega_n}, -\frac{\text{LambertW}\left(-1, \frac{1}{10}e^{(-1)}\right)+1}{\omega_n}$$

### Part (a)
> omega[n]:=4;
$$\omega_n := 4$$

> solve(eta=.1,t);
-0.2404446896, 0.9724300425

Thus, the time $t$ required for the mass to reach the position $x = 0.1x_0$ when $\omega_n = 4$ rad/s is $t = 0.972$ sec.

### Part (b)
> omega[n]:=2: eta1:=eta:

```
> omega[n]:=4: eta2:=eta:
> omega[n]:=8: eta3:=eta:
> p1:=plot([eta1,eta2,eta3],t=0..3,color=black,labels=["time (s)",""], title="non-dimensional
displacement"):
> t1:=textplot([.58,.8,"2 rad/s"]):
> t2:=textplot([.67,.4,"4 rad/s"]):
> t3:=textplot([.58,.15,"8 rad/s"]):
> display(p1,t1,t2,t3);
```

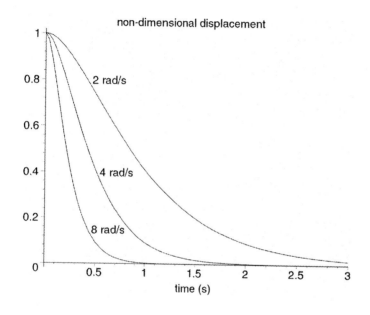

# 8.3 Sample Problem 8/6 (Forced Vibration of Particles)

The 100-lb piston is supported by a spring of modulus $k = 200$ lb/in. A dashpot of damping coefficient $c = 85$ lb-sec/ft acts in parallel with the spring. A fluctuating pressure $p = 0.625$ $\sin(30t)$ (psi) acts on the piston, whose top surface area is 80 in$^2$. Plot the response of the system for initial conditions $x_0 = 0.05$ ft and $\dot{x}_0 = 5, 0,$ and $-5$ ft/sec.

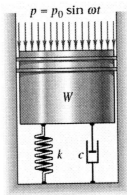

## Problem Formulation

The particular (steady state) solution was found in the sample problem in your text,

$$x_p = X \sin(\omega t - \phi)$$

where $X = 0.01938$ m, $\phi = 1.724$ rad and $\omega = 30$ rad/sec. Also from the sample problem, $\omega_n = \sqrt{k/m} = 27.8$ rad/sec and $\zeta = c/2m\omega_n = 0.492$.

The complete solution is found by adding the complementary (transient) and particular solutions. Since the system is underdamped ($\zeta < 1$), the complementary solution is,

$$x_c = Ce^{-\zeta\omega_n t} \sin(\omega_d t + \psi)$$

where $\omega_d = \omega_n \sqrt{1-\zeta^2} = 24.2$ rad/sec. The displacement of the system is

$$x = x_c + x_p = Ce^{-\zeta\omega_n t} \sin(\omega_d t + \psi) + X \sin(\omega t - \phi)$$

$$x = Ce^{-13.68t} \sin(24.2t + \psi) + 0.01938\sin(30t - 1.724)$$

The velocity is found by differentiating $x$,

$$\dot{x} = C\omega_d e^{-\zeta\omega_n t} \cos(\omega_d t + \psi) - C\zeta\omega_n e^{-\zeta\omega_n t} \sin(\omega_d t + \psi) + X\omega \cos(\omega t - \phi)$$

$$\dot{x} = Ce^{-13.68t}\left(24.2\cos(24.2t + \psi) - 13.68\sin(24.2t + \psi)\right) + 0.5814\cos(30t - 1.724)$$

The constants $C$ and $\psi$ are found from the initial conditions,

$$x_0 = 0.05 = C \sin \psi + X \sin \phi = C \sin \psi - 0.0192$$

$$\dot{x}_0 = C\omega_d \cos \psi - C\zeta\omega_n \sin \psi + X\omega \cos \phi = 24.2C \cos \psi - 13.7C \sin \psi - 0.0887$$

The first equation can be solved for $C = 0.0692/\sin \psi$. Substitution into the second equation gives $\psi$.

$$\psi = \tan^{-1}\left( \frac{1.675}{\dot{x}_0 + 1.035} \right) \qquad C = \frac{0.0692}{\sin \psi}$$

This yields the following values for $C$ and $\psi$.

$x_0 = 0.05$ ft, $\dot{x}_0 = 5$ ft/s, $C = 0.259$ ft, $\psi = 0.271$ rad
$x_0 = 0.05$ ft, $\dot{x}_0 = 0$ ft/s, $C = 0.081$ ft, $\psi = 1.017$ rad
$x_0 = 0.05$ ft, $\dot{x}_0 = -5$ ft/s, $C = -0.178$ ft, $\psi = -0.400$ rad

## *Maple Worksheet*

```
> restart; with(plots):
> x := C*sin(24.2*t+psi)*exp(-13.68*t)+.01938*sin(30*t-1.724);
```
$$x := C \sin(24.2\,t + \psi)\, e^{(-13.68\,t)} + .01938 \sin(30\,t - 1.724)$$

```
> psi:=arctan(1.675/(xd0+1.035));
```
$$\psi := \arctan\left( 1.675 \frac{1}{xd0 + 1.035} \right)$$

```
> C:=.0692/sin(psi);
```
$$C := .04131343284(xd0 + 1.035)\sqrt{1 + \frac{2.805625}{(xd0 + 1.035)^2}}$$

```
> xd0:=5: x1:=x: xd0:=0: x2:=x: xd0:=-5: x3:=x:
>
> p1:=plot([x1,x2,x3],t=0..0.5,color=black,labels=[`time (sec)`,`position (feet)`],labeldirections =
[HORIZONTAL,VERTICAL]):
> t1:=textplot([.082,.13,"5 ft/s"]):
> t2:=textplot([.04,.05,"0 ft/s"]):
> t3:=textplot([.095,-.07,"- 5 ft/s"]):
>
> display(p1,t1,t2,t3);
```

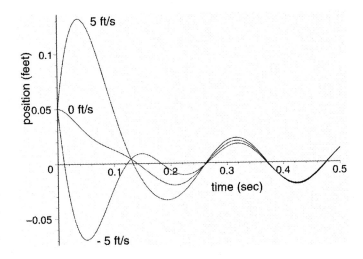

Notice how quickly the three cases converge to the steady state solution.

Printed in the United Kingdom
by Lightning Source UK Ltd.
118931UK00001B/89-90